Microscopy, Optical Spectroscopy, and Macroscopic Techniques

Methods in Molecular Biology

John M. Walker, Series Editor

22. **Microscopy, Optical Spectroscopy, and Macroscopic Techniques,** edited by ***Christopher Jones, Barbara Mulloy, and Adrian H. Thomas, 1994***
21. **Protocols in Molecular Parasitology,** edited by ***John E. Hyde, 1993***
20. **Protocols for Oligonucleotides and Analogs,** edited by ***Sudhir Agrawal, 1993***
19. **Biomembrane Protocols:** *I. Isolation and Analysis,* edited by ***John M. Graham and Joan A. Higgins, 1993***
18. **Transgenesis Techniques,** edited by ***David Murphy and David A. Carter, 1993***
17. **Spectroscopic Methods and Analyses,** edited by ***Christopher Jones, Barbara Mulloy, and Adrian H. Thomas, 1993***
16. **Enzymes of Molecular Biology,** edited by ***Michael M. Burrell, 1993***
15. **PCR Protocols,** edited by ***Bruce A. White, 1993***
14. **Glycoprotein Analysis in Biomedicine,** edited by ***Elizabeth F. Hounsell, 1993***
13. **Protocols in Molecular Neurobiology,** edited by ***Alan Longstaff and Patricia Revest, 1992***
12. **Pulsed-Field Gel Electrophoresis,** edited by ***Margit Burmeister and Levy Ulanovsky, 1992***
11. **Practical Protein Chromatography,** edited by ***Andrew Kenney and Susan Fowell, 1992***
10. **Immunochemical Protocols,** edited by ***Margaret M. Manson, 1992***
9. **Protocols in Human Molecular Genetics,** edited by ***Christopher G. Mathew, 1991***
8. **Practical Molecular Virology,** edited by ***Mary K. L. Collins, 1991***
7. **Gene Transfer and Expression Protocols,** edited by ***Edward J. Murray, 1991***
6. **Plant Cell and Tissue Culture,** edited by ***Jeffrey W. Pollard and John M. Walker, 1990***
5. **Animal Cell Culture,** edited by ***Jeffrey W. Pollard and John M. Walker, 1990***
4. **New Nucleic Acid Techniques,** edited by ***John M. Walker, 1988***
3. **New Protein Techniques,** edited by ***John M. Walker, 1988***
2. **Nucleic Acids,** edited by ***John M. Walker, 1984***
1. **Proteins,** edited by ***John M. Walker, 1984***

Methods in Molecular Biology • 22

Microscopy, Optical Spectroscopy, and Macroscopic Techniques

Edited by

Christopher Jones, Barbara Mulloy, and Adrian H. Thomas

National Institute for Biological Standards and Control, South Mimms, Potters Bar, UK

Humana Press Totowa, New Jersey

999 Riverview Drive, Suite 208
Totowa, New Jersey 07512

Printed in the United States of America. 9 8 7 6 5 4 3 2 1

Library of Congress Cataloging in Publication Data

Main entry under title:
Methods in molecular biology.

Microscopy, Optical spectroscopy, and macroscopic techniques / edited by
Christopher Jones, Barbara Mulloy, and Adrian H. Thomas.
p. cm. — (Methods in molecular biology ; 22)
Includes index.
ISBN 0-89603-232-9
1. Biomolecules—Spectra. 2. Electron microscopy. 3. Molecular biology—methodology. I. Jones, Christopher, 1954– . II. Mulloy, Barbara. III. Thomas, A. H. (Adrian H.) IV. Series: Methods in molecular biology (Totowa, NJ) ; 22.
QP519.9.S6M53 1993
578'.45—dc20 93-23977
CIP

Preface

This is the second of three volumes of *Methods in Molecular Biology* that deal with Physical Methods of Analysis. The first of these, *Spectroscopic Methods and Analyses* dealt with NMR spectroscopy, mass spectrometry, and metalloprotein techniques, and the third will cover X-ray crystallographic methods.

As with the first volume, *Microscopy, Optical Spectroscopy, and Macroscopic Techniques* is intended to provide a basic understanding for the biochemist or biologist who needs to collaborate with specialists in applying the techniques of modern physical chemistry to biological macromolecules.

The methods treated in this book fall into four groups. Part One covers microscopy, which aims to visualize individual molecules or complexes of several molecules. Electron microscopy is the more familiar of these, while scanning tunneling microscopy is a new and rapidly developing tool. Methods for determining the shapes and sizes of molecules in solution are described in Part Two, which includes chapters on X-ray and neutron scattering, light scattering, and ultracentrifugation. Calorimetry, described in Part Three, provides the means to monitor processes involving thermodynamic changes, whether these are intramolecular, such as conformational transition, or the interactions between solutes or between a solute and its solvent. Part Four is concerned with optical and infrared spectroscopy and describes applications ranging from the measurement of protein concentration by UV absorbance to the analysis of secondary structure using circular dichroism and Fourier-transform infrared spectroscopy.

As in our previous volume, authors were asked to emphasize those practical aspects of these techniques that reflect on their application, such as constraints on sample quantity, purity, and presentation, the level of time and expense involved, the problems the technique is best suited to solve, and how the results may be interpreted. We hope that each of these volumes will encourage effective and productive multidisciplinary approaches to problems in molecular biology and biochemistry.

We would particularly like to thank Robin Wait for his continuing help in the preparation of this book.

Christopher Jones
Barbara Mulloy
Adrian H. Thomas

Contents

Contributors

DENNIS CHAPMAN • *Department of Protein and Molecular Biology, Royal Free Hospital School of Medicine, University of London, UK*

MANO S. CHEEMA • *Department of Pharmaceutical Sciences, University of Nottingham, UK*

ALAN COOPER • *Department of Chemistry, Glasgow University, Glasgow, Scotland, UK*

MARTYN C. DAVIES • *Department of Pharmaceutical Sciences, University of Nottingham, UK*

ALEX F. DRAKE • *Department of Chemistry, Birkbeck College, London, UK*

CARLA W. GRAY • *Program in Molecular and Cell Biology, The University of Texas at Dallas, Richardson, TX*

STEPHEN E. HARDING • *Department of Applied Biochemistry and Food Science, University of Nottingham, UK*

PARVEZ I. HARIS • *Department of Protein and Molecular Biology, Royal Free Hospital School of Medicine, University of London, UK*

DAVID E. JACKSON • *Department of Pharmaceutical Sciences, University of Nottingham, UK*

CHRISTOPHER M. JOHNSON • *Department of Chemistry, Glasgow University, Glasgow, Scotland, UK*

STEPHEN J. PERKINS • *Departments of Biochemistry and Chemistry, and of Protein and Molecular Biology, Royal Free Hospital School of Medicine, London, UK*

SAUL J. B. TENDLER • *Department of Pharmaceutical Sciences, University of Nottingham, UK*

PAUL G. VARLEY • *National Institute for Biological Standards and Control, South Mimms, Potters Bar, UK*

PHILLIP M. WILLIAMS • *Department of Pharmaceutical Sciences, University of Nottingham, UK*

PART I

MICROSCOPY

CHAPTER 1

Electron Microscopy of Protein–Nucleic Acid Complexes

Enhanced High-Resolution Shadowing

Carla W. Gray

1. Introduction

I describe herein a method in which pretreatment with low concentrations of uranyl acetate is used to increase the structural rigidity of protein–nucleic acid complexes, thereby substantially enhancing the information content of images obtained by high-resolution shadowcasting with tungsten. The visualization of three-dimensional objects by heavy-metal shadowcasting is a long-established technique in electron microscopy *(1)*, and it has long been realized that the highest resolution could be achieved by using carbon–metal mixtures or high-melting-point metals for the evaporation *(2,3)*. Double-stranded DNAs, 2 nm in diameter, are readily visualized by high-resolution shadowing with tungsten *(4)*.

Protein–DNA complexes formed in vitro by mixing the M13 or fd gene 5 single-stranded DNA binding protein* with single-stranded DNA were shown by negative staining in an early study *(5)* to be apparently helical, with an estimated helix diameter of 10 nm and an approx 6.5–7 nm distance between helical turns. Such dimensionally

*The M13 and fd gene 5 proteins are single-stranded DNA binding proteins of identical amino acid sequence encoded by the closely related M13 and fd strains of filamentous bacterial viruses.

From: *Methods in Molecular Biology, Vol. 22: Microscopy, Optical Spectroscopy, and Macroscopic Techniques* Edited by: C. Jones, B. Mulloy, and A. H. Thomas

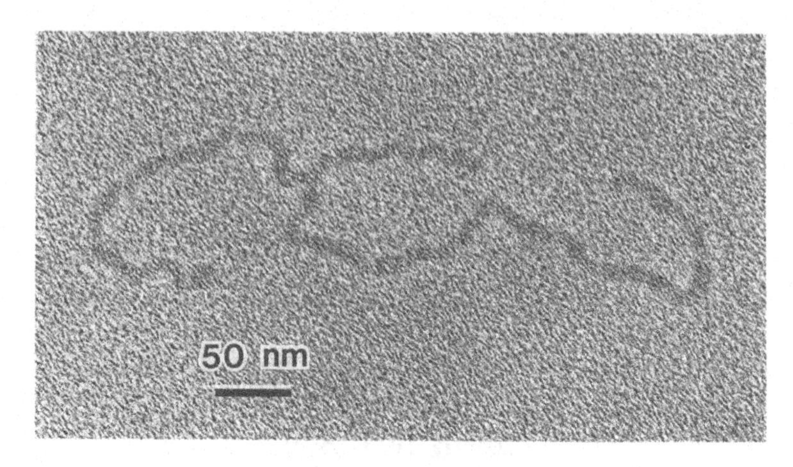

Fig. 1. Complex of fd gene 5 protein with viral DNA, formed by mixing the protein and DNA in vitro. The complex was adsorbed from 0.01*M* ammonium acetate (pH 6.9) onto a glow-discharge-activated carbon, dehydrated by passage through graded ethanol solutions before air-drying *(6)*, and rotary-shadowed with tungsten. The same complexes rinsed with water rather than ethanol have essentially the same appearance as shown here, as is also true for fd complexes isolated from infected cells and prepared as described here with drying from water or ethanol. Scale bar = 50 nm.

large structural features should be resolved by tungsten shadowing. It was quite surprising, therefore, to find that complexes of M13 viral DNA with the M13 gene 5 protein, isolated from virus-infected bacterial cells, did not show the expected three-dimensional structure when shadowed with tungsten *(6)*. This finding led to the conclusion (which was later shown to be incorrect *[7,8]*) that complexes formed in vivo did not have the same structure as had been shown by negative staining for complexes reconstituted in vitro.

Why did tungsten shadowing fail to delineate the helical structure in the case just cited? Poor structural definition is also obtained in my laboratory upon tungsten shadowing of in vivo and in vitro complexes of the fd gene 5 protein with fd viral DNA, using methods similar to those used in *(6)*; an example is shown in Fig. 1. Although the shadowed complex in the figure has roughly the expected overall dimensions (length and width), the individual helical turns cannot be distinguished, and the detailed structure is irregular. A possible reason for the lack of structural definition was revealed by our recent studies

(8) using uranyl acetate staining and specimen tilting to elucidate the three-dimensional structures of the fd complexes. In those studies, it was found that considerable flattening of the complexes occurs upon adsorption to glow-discharge-activated carbon; calculations indicated that the apparent diameter of 9.5–10.5 nm for the flattened helices should be corrected to a diameter of 7–9 nm before flattening.

Flattening of a specimen will of course reduce the resolution of its structural features by shadowcasting, which depends on the elevation of discrete structures above the supporting surface and on their spatial separation from neighboring structures. Moreover, the uranyl acetate stain used in the tilting experiments is itself believed to provide structural support for specimens partially embedded in the stain *(9,10)*. Thus, it seemed likely that even more pronounced flattening of the M13 or fd protein–DNA complexes could have occurred in the preparations used for shadowcasting, in which the specimens were dried from aqueous or ethanolic solutions in the absence of uranyl acetate.

I therefore set about to see whether it would be possible to develop a technique combining high-resolution tungsten shadowing of nucleoprotein complexes with the use of uranyl acetate stain as a medium for specimen support. Enhanced contrast has previously been observed for DNAs stained with uranyl acetate prior to shadowing *(11,12)*. Surprisingly, I found that pretreatment with uranyl acetate provided substantial support and greatly improved structural resolution in shadowed nucleoprotein complexes, *even at concentrations below those required for the uranyl salts to be visible as a stain.* The specimen contrast in Figs. 2–4 is entirely owing to the tungsten deposits and not to the uranyl acetate, as was determined in tests that are described together with the method.

2. Materials

1. The protein–DNA complexes to be visualized: These must be freshly purified, and free of contaminating proteins, lipids, oils, salts, other nonvolatile components, detergents, and nonaqueous solvents. There must be no contaminating nucleases or proteases. We frequently find it necessary to repurify proteins and nucleic acids obtained commercially or from other laboratories. Simple dialysis may suffice to remove low-mol-wt contaminants, or ethanol precipitation (of nucleic acids) or gel filtration may be used. It is preferable to mix the DNAs and proteins just prior to mounting them for microscopy; the proteins should not be

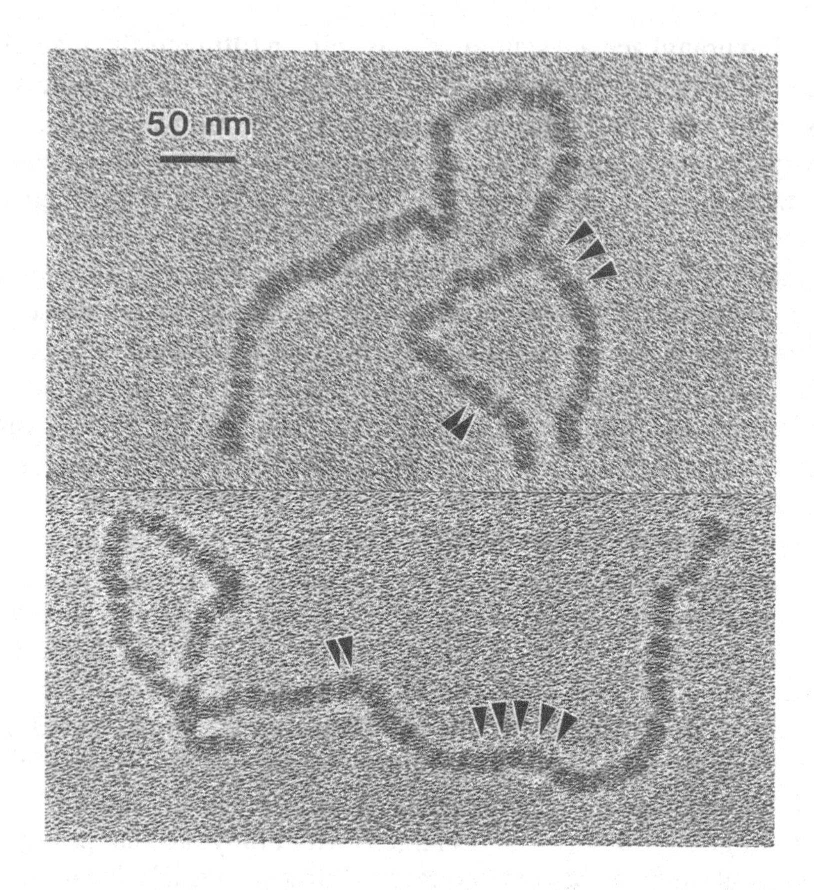

Fig. 2. Complexes of fd gene 5 protein with viral DNA. The complexes were isolated from infected cells, adsorbed to a glow-discharge-activated carbon, rinsed with water, treated with 0.5% uranyl acetate, and rotary-shadowed with tungsten. The arrows point to individual turns of the nucleoprotein helix that are visible after uranyl acetate pretreatment. Scale bar = 50 nm.

present in such excess quantities that they will contribute significantly to the background. About 0.2–2 μmol of DNA (measured as the concentration of nucleotide phosphates), together with protein added at an appropriate ratio, will be needed in each 50-μL mixture used for adsorption of the protein–DNA complexes to a single specimen grid.

2. Purified water: Water is distilled and then deionized in a "Milli-Q" system (Millipore Corp., Bedford, MA) consisting of one cellulose ester prefilter cartridge, two ion-exchange cartridges, and a 0.22-μm filter, in series. No activated charcoal filter is included in our system, because of a tendency of this filter to release minute charcoal particles. The

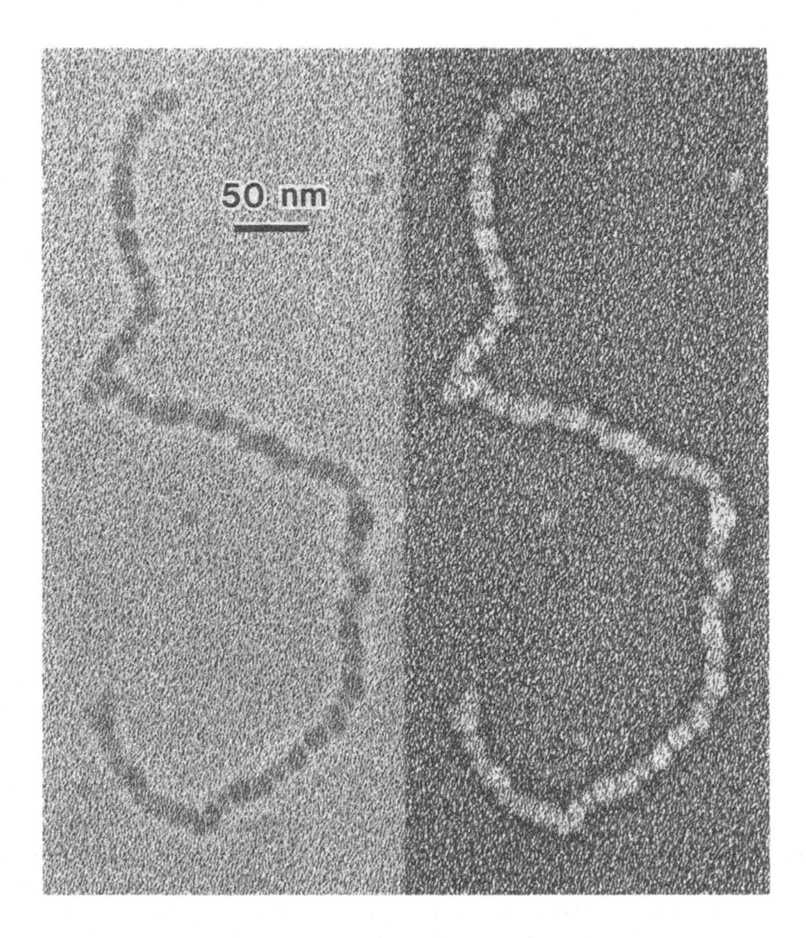

Fig. 3. A complex prepared as for Fig. 2, printed directly from the electron-image negative (left) or from a copy negative having reversed contrast (right). Note the dark shadow around the complex on the right, corresponding to a light halo seen before contrast reversal. The shadow demonstrates that the structure is three-dimensional. Scale bar = 50 nm.

Milli-Q system is constructed of highly inert, noneluting materials, and we find that the water (18 megΩ/cm as dispensed) can be used for most procedures in electron microscopy. Alternatively, we sometimes use tap distilled water that has been redistilled through a series of two 24-in. borosilicate glass Vigreux columns.

3. Buffers for mixing of the protein and DNA to form complexes: The use of a minimal number of buffer components and a low concentration of salts (<0.05*M*) is desirable, since salts and some other buffer compo-

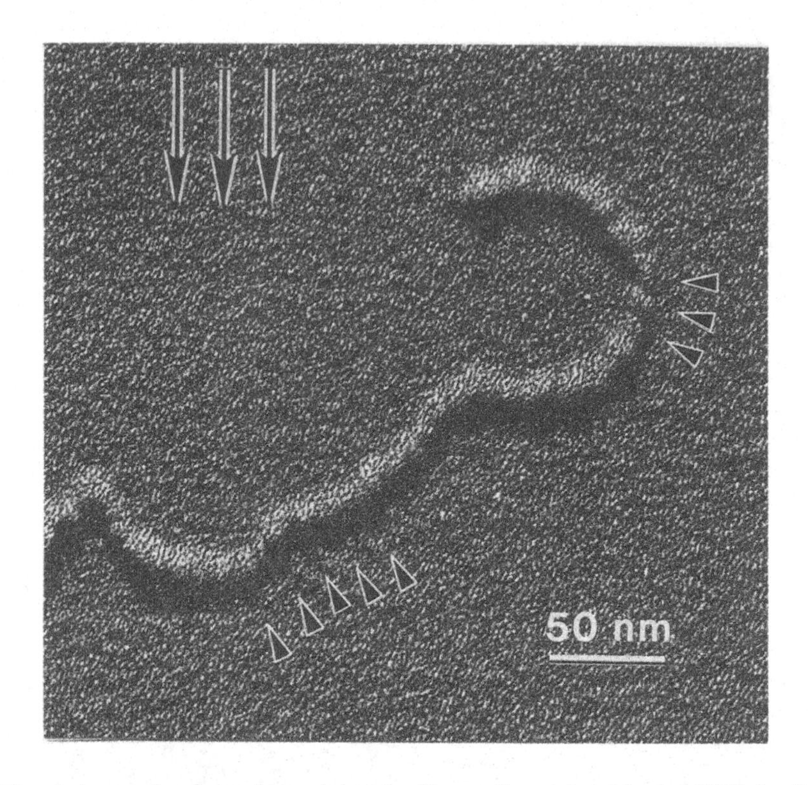

Fig. 4. A complex formed by mixing the fd gene 5 protein with viral DNA in vitro. The complex was prepared for electron microscopy as described in the legend to Fig. 2, except that the complex shown here was unidirectionally shadowed and was then printed with contrast reversal. The arrows at the top of the figure indicate the direction of tungsten shadowing. The arrows to the right indicate resolved helical turns (helix axis parallel to the direction of shadowing). The arrows below the complex point to the shadows formed by individual turns of the helix. Scale bar = 50 nm.

nents may interfere with specimen adsorption or visualization. (For example, glycerol or sucrose leaves a nonvolatile residue after drying.) Stocks of 1*M* buffer should be filtered through a 0.22-µm fiberless polycarbonate filter to remove particulates that can interfere with mounting for electron microscopy. The buffer stocks should then be stored at 4°C in tightly closed borosilicate glass bottles having caps lined with Teflon™ or inert plastic.

4. Glutaraldehyde: This reagent should be highly purified for electron microscopy (e.g., vacuum distilled), polymer-free, supplied in sealed ampules under inert gas and stored at –20°C. The contents of one ampule are diluted to 8% in purified water and stored at –20°C in a tightly

capped borosilicate glass tube with a Teflon™-lined screw cap; this solution can be used for as long as 6 mo.

5. Uranyl acetate, analytical reagent grade: A 0.5–1% (w/v) solution in purified water is dissolved by stirring for 30 min in a borosilicate glass beaker. The solution is tightly covered with wax film and is used within a few hours. Uranyl acetate is weakly radioactive and should be handled with appropriate caution; discarded solutions should be collected and properly disposed of as radioactive waste.
6. Carbon films: Clean carbon films 8–10 nm thick, on 500-mesh copper grids with tabs (handles), are made in an Edwards E306A evaporator equipped with a quartz crystal film thickness monitor and a liquid nitrogen trap, evacuating just to 1×10^{-2} Pa, then immediately burning off contaminants (shutter closed, carbon rods brought to a red glow), and finally evaporating for several seconds from high purity carbon rods milled to form 1-mm tips. The films are evaporated onto freshly cleaved mica, floated from the mica onto purified water in a recessed solid Teflon™ dish, picked up on copper grids, and dried for 30 min under a lamp.
7. Tungsten source: Tungsten rods are used in the Edwards twin electron beam source. Alternatively, tungsten wire of 0.5 mm diameter may be evaporated by direct resistance heating in an evaporator fitted with a 2 kVA transformer and heavy-duty conduits to handle a current of 30–35 A at 45–50 V, which will be sufficient to evaporate the tungsten. We used an electron beam source for the experiments shown in Figs. 1–4, because it provides a point source of evaporated tungsten, which should give optimal resolution when shadowing unidirectionally (Fig. 4).

3. Methods

Stock solutions of DNA and protein, and DNA–protein mixtures, are prepared in high-quality, contamination-free, noncolored 0.5- or 1.5-mL conical polypropylene tubes. Pipeting is done using air-displacement microliter pipeters with nonwetting, smoothly molded polypropylene tips of high quality. We prefer tips with beveled (not blunt) tip ends for precise pipeting of small volumes to make DNA–protein mixtures. Care must be taken to avoid touching the pipet tips or tubes in order to avoid contamination with oils and nucleases from the skin.

1. Prepare the DNA–protein mixtures. A vol of 50 µL will be required for each specimen grid. Preparation of enough material for four grids is recommended, so that more than one attempt can be made at shadowing, both rotary and unidirectional. The final concentration needed will be equivalent to about 4–40 mµmol of DNA phosphate/mL, but the

required concentration will vary with the adsorptive properties of the DNA and protein in a buffer of given composition, pH, and ionic strength. The correct concentrations can only be determined experimentally.

2. Clean a pure, white Teflon™ surface by rubbing it with ethanol (reagent grade, 95% ethanol, approx 5% H_2O) and then rinsing with purified water. For each specimen grid, set up a row of droplets on the Teflon™: an empty space for one droplet, then two 50-µL droplets of purified water, then one 50-µL droplet of 0.5% uranyl acetate.
3. Initiate glutaraldehyde fixation. This will be required in most cases to maintain noncovalent protein–DNA associations during adsorption to the charged (glow-discharge-activated) carbon film. (To confirm that fixation is effective and yields an unperturbed structure, stained preparations can be made by the method of Valentine et al. *[13]*, which does not require fixation.) Add 0.5 µL of 8% glutaraldehyde (or 2 µL of 2% glutaraldehyde) to the bottom of a clean 0.5-mL conical polyethylene tube. Immediately add 50 µL of the DNA–protein mixture, and gently mix by pipeting up and down once with a pipeter adjusted to 50 µL. Incubate the reaction at ambient temperature (20–25°C) for 20 min.
4. Meanwhile, place an appropriate number of carbon-coated grids, carbon side up, on an inverted Petri dish (precleaned with 95% ethanol and dried); subject the grids to glow discharge. For this purpose, we use two L-shaped aluminum rods (6.5 mm in diameter) fitted to the high-tension electrodes of the Edwards E306A evaporator. Glow discharge is carried out at 10–20 Pa (only the rotary pump is in operation), with the grids placed on the Petri dish directly below the parallel, 15 cm long horizontal segments of the two aluminum rods at a distance of 4 cm from the rods. The discharge, at 40% of maximum voltage (i.e., at approx 2 kV), is continued for 50 s, and the grids are used for adsorptions within 10 min thereafter.
5. For each DNA–protein mixture to be examined, place a 50-µL droplet of the mixture on the Teflon™ surface, at the beginning of a row of droplets prepared as described in item 2 of this section. Pick up a grid by its handle and touch it, carbon side down, to the top of the droplet containing the DNA–protein mixture for 30–60 s, and then wick off excess solution from the grid onto a filter paper (the grid is held perpendicular to the flat surface of the filter paper). Next, touch the grid to each of the two water droplets for 1 s each, and finally touch the grid to the uranyl acetate droplet for 20 s, wicking off excess liquid after each step.
6. Dry the grid for 10 s by holding it within 2–3 cm of a lamp bulb (we use an illuminator having a 30-W bulb and a polished metal reflector), and

then dry it for 10 min, carbon side up, on a filter paper that is under the lamp and about 12 cm from the bulb.

7. The grids should now be examined to check that the uranyl acetate staining is not too strong or weak. This should be done in an electron microscope fitted with a liquid nitrogen anticontaminator, to avoid contaminating the specimen prior to shadowing it. Suitable settings are 60 kV, with a 60-μm objective aperture and an electron-optical magnification of about 25,000×. We use an objective-lens focal length of 2.8 mm in a Carl Zeiss EM10CA transmission electron microscope.

   Most areas of the grid should be featureless, indicating that the staining has been very light. Within each square framed by the copper mesh of the grid, one or more areas may show some staining of the DNA–protein complexes. These stained areas are usually in the same relative location (for example, toward the upper right corner) in most squares of the grid. If the visibly (lightly) stained areas occupy <1/3 to 1/2 of the area of each grid square, the grid is suitable for shadowing. Note the locations of the stained areas with respect to the asymmetric marker at the grid center, for later reference. If no stained areas are seen, the staining may or may not be too light; shadowing of the grid should be tried. If adjustment of the stain intensity is needed, try doubling or halving the concentration of the uranyl acetate solution.

8. Shadowing with tungsten should be done within a few hours before electron micrographs are to be exposed, since diffusion of the deposited tungsten can occur rapidly. We use an electron beam source, with the 3-mm grids held very lightly by two parallel strips of double-stick tape placed 2.9 mm apart on a rotating disk at an angle of 20° to the source. An Edwards FTM5 quartz crystal film thickness monitor (capable of measuring in increments of 0.1 nm) is used to measure a thickness of 1.5 nm for the deposited tungsten.

   After shadowing, the areas of the grid on which there was visible uranyl acetate stain will have a distinctly different appearance from those areas that did not. Use of the reference marker as described in item 7 of this section will make it possible to identify unambiguously the areas of the grid that were *not* visibly stained, in which the now visible images will be purely the result of tungsten shadowing. The nonstained areas are those that are of interest for the method described here. Examine a number of these areas, throughout the grid; the contrast may be better in some regions of the grid than in others. Occasionally, the shadowing may fail, giving little or no contrast; in that case, try shadowing another grid on which the same specimen has been mounted.

The results obtained with this method are illustrated in Figs. 2–4. Figure 2 shows two examples of the same type of helical complex as was shown in Fig. 1. Rotary shadowing with tungsten was used to visualize the complexes shown in both Figs. 1 and 2. The essential difference between the preparations shown in these two figures is that the complexes in Fig. 2 were treated with uranyl acetate prior to drying, whereas the complex in Fig. 1 was not. *The individual helical turns are visible* in the complexes in Fig. 2 (arrows), presumably because of support of the fragile helical coils by invisibly small quantities of uranyl acetate associated with the DNA–protein complexes. Sometimes two turns are close together, appearing to have a double width (double arrows). The dimensions of the helical turns are in close agreement with the dimensions obtained in our earlier studies of the same complexes embedded in negative stain *(7,8)*. The complexes in Fig. 2 have a greater apparent diameter than does that in Fig. 1, because the supported helical turns are elevated above the surface of the carbon substrate and therefore receive heavier deposits of tungsten.

We should point out that the prints in Fig. 2 are made directly from the electron-image negatives. The projecting helical turns are *dark* because of tungsten deposits, whereas the narrow spaces between the turns are shielded from receiving deposits and remain *light*. Opposite effects would be observed if we were seeing effects owing to staining by the uranyl acetate, which collects in the crevices and makes them dark, while leaving the protruding helix turns unstained.

Figure 3 shows another complex prepared using uranyl acetate pretreatment. The image on the left was produced directly from the electron-image negative, whereas that on the right was produced from a second negative made as a contrast-reversed, contact copy of the first. The resulting contrast-reversed image is more typically used to represent shadowed preparations, because it shows the shadows as being dark, as one intuitively expects them to be. There is a shadow (dark halo) surrounding the complex in the righthand image, corresponding to a light halo in the lefthand image. There is no light halo around the complex in Fig. 1, confirming that it is indeed nearly flat in the absence of uranyl acetate pretreatment.

Finally, Fig. 4 shows a uranyl–acetate pretreated complex that has been *unidirectionally* shadowed with tungsten, using a point source for highest resolution. The complex in this case takes on an appearance that gives the viewer the best intuitive perception of its three-dimensional structure, with shadows that emulate lighting from one direction as we are accustomed to seeing. It is more difficult to resolve the individual helical turns using unidirectional metal shadowing, but clear resolution does occur in some instances when the helix axis is oriented parallel to the direction of shadowing (arrows at upper right). Moreover, a careful examination of the *shadows* below the complex (arrows) very clearly reveals a corrugated structure corresponding to the neighboring turns of the helix: This is the clearest, most direct demonstration of the actual structure in three dimensions.

Acknowledgments

The author wishes to acknowledge excellent technical assistance by B. A. Davis. This work was supported by National Institutes of Health Research Grant 5-RO1-GM34293-03 (to C. W. G.), by NIH Biomedical Research Support Grant 2S07-RR07133-21, by NIH Small Instrumentation Grant 1-S15-NS25421-01, and by National Science Foundation Instrumentation Grant PCM-8116109.

References

1. Williams, R. C. and Wycoff, R. W. G. (1946) Applications of metallic shadow-casting to microscopy. *J. Appl. Phys.* **17,** 23.
2. Bradley, D. E. (1959) High-resolution shadow-casting technique for the electron microscope using the simultaneous evaporation of Pt and carbon. *Br. J. Appl. Phys.* **10,** 198.
3. Abermann, R., Salpeter, M. M., and Bachmann, L. (1972) High resolution shadowing, in *Principles and Techniques of Electron Microscopy,* vol. 2 (Hayat, M. A., ed.), Van Nostrand Reinhold, New York, pp. 197–217.
4. Griffith, J., Huberman, J. A., and Kornberg, A. (1971) Electron microscopy of DNA polymerase bound to DNA. *J. Mol. Biol.* **55,** 209–214.
5. Alberts, B., Frey, L., and Delius, H. (1972) Isolation and characterization of gene 5 protein of filamentous bacterial viruses. *J. Mol. Biol.* **68,** 139–152.
6. Pratt, D., Laws, P., and Griffith, J. (1974) Complex of bacteriophage M13 single stranded DNA and gene 5 protein. *J. Mol. Biol.* **82,** 425–439.
7. Gray, C. W., Kneale, G. G., Leonard, K. R., Siegrist, H., and Marvin, D. A. (1982) A nucleoprotein complex in bacteria infected with Pf1 filamentous virus: identification and electron microscopic analysis. *Virology* **116,** 40–52.
8. Gray, C. W. (1989) Three-dimensional structure of complexes of single-stranded DNA-binding proteins with DNA. IKe and fd gene 5 proteins form left-handed helices with single-stranded DNA. *J. Mol. Biol.* **208,** 57–64.
9. Haschemeyer, R. H. and Myers, R. J. (1972) Negative staining, in *Principles and Techniques of Electron Microscopy,* vol. 2 (Hayat, M. A., ed.), Van Nostrand Reinhold, New York, pp. 101–147.

10. Akey, C. W. and Edelstein, S. J. (1983) Equivalence of the projected structure of thin catalase crystals preserved for electron microscopy by negative stain, glucose, or embedding in the presence of tannic acid. *J. Mol. Biol.* **163,** 575–612.
11. Vollenweider, J. J., Sogo, J. M., and Koller, Th. (1975) A routine method for protein-free spreading of double- and single-stranded nucleic acid molecules. *Proc. Natl. Acad. Sci. USA* **72,** 83–87.
12. Williams, R. C. (1977) Use of polylysine for adsorption of nucleic acids and enzymes to electron microscope specimen films. *Proc. Natl. Acad. Sci. USA* **74,** 2311–2315.
13. Valentine, R. C., Shapiro, B. M., and Stadtman, E. R. (1968) Regulation of glutamine synthetase, XII. Electron microscopy of the enzyme from *Escherichia coli. Biochemistry* **7,** 2143–2152.

CHAPTER 2

Visualization of Unshadowed DNA by Electron Microscopy

Adsorption to a Bacitracin Film

Carla W. Gray

1. Introduction

I describe herein a simple, rapid method in which double-stranded DNAs are adsorbed with little or no distortion to a film of the oligopeptide bacitracin on a glow-discharge-activated carbon. The DNAs can be readily visualized by positive staining with uranyl acetate, which forms a smooth trace along the DNA contour against a very even background. This eliminates the need for heavy-metal shadowing, which would thicken the DNA and obscure details of the actual path traced by the DNA double helix. The stained specimen can subsequently be shadowed with tungsten if desired, however, to confirm the positioning of proteins bound to the DNA.

The ability to visualize configurational arrangements of biological nucleic acids has long been of interest. The supercoiling and sequence-specified bending of double-stranded DNAs, the winding of DNA around nucleosomes, RNA splicing, and the 5'-linkage of paired retrovirus genomic RNAs are all phenomena for which visualization by electron microscopy has provided revealing first insights or essential confirmation. Electron microscopy will continue to be important in defining higher-order interactions that may not always be easily

From: *Methods in Molecular Biology, Vol. 22: Microscopy, Optical Spectroscopy, and Macroscopic Techniques* Edited by: C. Jones, B. Mulloy, and A. H. Thomas

observed by other means. An example in which there is growing interest is gene regulation through interactions between distant sites on a DNA (separated by hundreds of base pairs): These phenomena should be detectable by virtue of the formation of a DNA loop between the interacting loci.

The first landmark in nucleic acid visualization was the development of the cytochrome *c* spreading technique *(1)* by Kleinschmidt and collaborators. This was followed by important modifications to achieve spreading in droplets *(2)* or to spread single-stranded DNAs in the presence of formamide *(3)*. The cytochrome *c* spreading technique depends on the surfactant properties of the cytochrome *c* protein, which forms a surface monolayer at the meniscus of a solution; DNAs mixed with the cytochrome *c* become embedded in the protein monolayer. The DNAs become heavily coated with the cytochrome *c* protein in the process, so that after adsorption of the monolayer to a support film and heavy-metal deposition by rotary shadowing, the apparent diameter of a double-stranded DNA helix is about five times its true 2-nm value. This of course obscures local curvatures in the DNA, and it obscures noncytochrome proteins bound to the DNA. The thickening of the DNA is reduced by a method using a small-molecule surfactant in place of the cytochrome protein *(4)*, but one still has to deal with the difficulty of achieving proper formation of the surfactant film at a meniscus, and heavy-metal shadowing is still used for visualization.

Griffith *(5)* has visualized DNAs adsorbed directly to glow-discharge-activated carbon films, using tungsten shadowing and sometimes tilting the specimen to achieve good contrast. Here again, the heavy-metal deposits add substantially to specimen thickness and tend to obscure detail. Williams *(6)* used tungsten shadowing to visualize DNAs adsorbed to a polylysine film. I have adsorbed double-stranded DNAs to activated carbon and have directly visualized them at their true 2 nm diameter using a very thin, uniform layer of uranyl acetate as a negative stain *(7)*. The DNA is made visible by virtue of the fact that the stain is excluded from the volume occupied by the DNA, and the diameter of the DNA is not artificially increased. The contrast in such micrographs is quite low, however, making it difficult to reproduce them for publication. (A good reproduction is shown in ref. *7.)* Also, I find that long DNA molecules (nicked circular DNA from PM2 virus, approx 7900 bp), which are adsorbed directly to glow-discharge-activated carbons, show evidence of localized desorption, stretching, and backfolding, which will introduce errors into length measurements.

Gregory and Pirie *(8,9)* have provided an excellent description of the use of the oligopeptide bacitracin (mol wt 1411 Da) as a wetting agent to aid in the adsorption of proteinaceous specimens to carbon films that have not been activated by glow discharge. Their technique involves drying sufficient bacitracin onto a grid to form a uniform monolayer. Large proteins mixed with the bacitracin solution are adsorbed in a uniform fashion, and can be visualized by metal shadowing or negative staining, with minimal interference from the finely granular bacitracin background. It occurred to this author that bacitracin might be similarly used to improve the adsorption of double-stranded DNAs to glow-discharge-activated carbon films. The following is a description of a simple method that achieves that objective. The DNA is *positively* stained at high contrast with uranyl acetate, allowing visualization of the DNA contours at true molecular dimensions. We find that long DNAs are more firmly and uniformly adsorbed to the bacitracin film than to a bare, glow-discharge-activated carbon.

2. Materials

1. Purified water: Water is distilled and then deionized in a "Milli-Q" system (Millipore Corp., Bedford, MA) consisting of one cellulose ester prefilter cartridge, two ion-exchange cartridges, and a 0.22-µm filter, in series; the water is 18 megΩ as dispensed. No activated charcoal filter is included in our system because of a tendency of this filter to release minute charcoal particles. We find that water from this system can be used for most procedures in electron microscopy. Alternatively, we sometimes use tap distilled water, which we have twice redistilled through a series of two 24-in. borosilicate glass Vigreux columns.
2. Carbon films: Clean carbon films 7–9 nm thick, on 500-mesh copper grids having tabs (handles), are made in an Edwards E306A evaporator equipped with a liquid nitrogen trap and an Edwards FTM5 quartz crystal film thickness monitor capable of measuring thickness in increments of 0.1 nm. We evacuate just to 1×10^{-2} Pa, then immediately burn off contaminants (shutter closed, carbon rods brought to a red glow), and finally evaporate for several seconds from high-purity carbon rods milled to form 1-mm tips. The films are evaporated onto freshly cleaved mica, floated from the mica onto purified water in a recessed solid Teflon™ dish, picked up on copper grids, and dried for 30 min under a lamp.

3. Bacitracin (Sigma Chemical Co., St. Louis, MO, 65,000 U/g): Bacitracin is harmful if absorbed through the skin, and it may cause allergic skin reactions. The powder is hygroscopic and should be stored desiccated at 0–5°C. We prepare a stock solution at 10 mg/mL in 10 m*M* ammonium acetate, pH 6.7– 6.9, and store it frozen at –20°C.
4. Ammonium acetate, analytical reagent grade: A 1*M* stock in purified water is filtered through a 0.22-µm fiberless polycarbonate filter and stored at 4°C in a tightly closed borosilicate glass bottle with a cap liner of Teflon™ or inert plastic. The pH of the solution (diluted to 0.05*M* for measurement) is generally 6.7–6.9.
5. Uranyl acetate, analytical reagent grade: A 2% (w/v) solution in purified water is dissolved by stirring for 30 min in a borosilicate glass beaker. The beaker is sealed with a wax film, and the solution is used within a few hours. Uranyl acetate is weakly radioactive and should be handled with appropriate caution; discarded solutions should be collected and properly disposed of as radioactive waste.
6. The double-stranded DNA to be visualized: The DNA must be purified and free of contaminating proteins, lipids, oils, salts, other nonvolatile components, detergents, and nonaqueous solvents, and of course it must be free of unwanted nucleases. We frequently find it necessary to repurify DNAs obtained commercially or from other laboratories. Simple dialysis may suffice to remove low-mol-wt contaminants, or ethanol precipitation or gel filtration may be used. DNA-binding proteins may be mixed with the DNA, but preferably not in excess quantities such that the proteins will contribute significantly to the background.

3. Methods

Once the stock solutions and materials have been prepared as described above, the execution of the method itself is rapid. Stock solutions and dilutions of bacitracin and DNA are prepared in high-quality, contamination-free, noncolored 0.5- or 1.5-mL conical polypropylene tubes. Pipeting is done using air-displacement microliter pipeters and smooth-surfaced, nonwetting polypropylene tips of high quality. We prefer the tips with beveled (not blunt) tip ends for precise pipeting of small volumes. Care must be taken to avoid touching the pipet tips or tubes in order to avoid contamination with oils and nucleases from the skin.

1. Dilute 5 µL of bacitracin (10 mg/mL) into 1.0 mL of 1 m*M* ammonium acetate, pH 6.7–6.9, to yield a solution containing 50 µg of the oligopeptide/mL.

2. Dilute the DNA stock to approx 2.5 μg/mL in 1 m*M* ammonium acetate, pH 6.7–6.9. Allow this dilution to stand for 5–10 min, in order for the DNAs to become disentangled from one another.
3. If DNA–protein complexes are to be examined, glutaraldehyde fixation will usually be required. *See* Chapter 8, this vol., for details.
4. Clean a pure, white Teflon™ surface by rubbing it with ethanol (reagent grade, 95% ethanol, approx 5% H_2O) and then rinsing with purified water. For each preparation to be examined, set up a row of droplets on the Teflon™: empty spaces for two droplets, then two 50-μL droplets of purified water, then one 50-μL droplet of 2% uranyl acetate.
5. Place an appropriate number of carbon-coated grids, carbon side up, on an inverted Petri dish (precleaned with 95% ethanol and dried); subject the grids to glow discharge. For this purpose, we use two L-shaped aluminum rods (6.5 mm in diameter) fitted to the high-tension electrodes of the Edwards E306A evaporator. Glow discharge is carried out at 10–20 Pa (only the rotary pump is in operation), with the grids placed on the Petri dish directly below the parallel, 15 cm long horizontal segments of the two aluminum rods at a distance of 4 cm from the rods. The discharge, at 40% of maximum voltage (i.e., at approx 2 kV), is continued for 50 s, and the grids are used for adsorptions within 10 min thereafter.
6. For each DNA preparation to be examined, place a 50-μL droplet of diluted bacitracin on the Teflon™ surface, followed by a 50-μL droplet of the DNA preparation, at the beginning of one of the rows of droplets described in step 4 above. Pick up a grid by its handle, touch it (carbon side down) to the top of the bacitracin droplet for 30 s, and then wick off excess solution from the grid onto a filter paper (the grid is held perpendicular to the flat surface of the filter paper). Next touch the grid to the DNA-containing droplet for 30 s, then to each of the two water droplets for 1 s each, and finally to the uranyl acetate droplet for 20–40 s, wicking off excess liquid after each step.
7. Dry the grid for 10 s by holding it within 2–3 cm of a lamp bulb (we use an illuminator having a 30-W bulb and a polished metal reflector), and then dry it for 10 min, carbon side up, on a filter paper that is under the lamp and about 12 cm from the bulb.
8. The specimens should be examined in a transmission electron microscope fitted with a liquid nitrogen anticontaminator, at 60 kV, using a 60-μm objective aperture and an electron-optical magnification of about 25,000×. We use a Carl Zeiss EM10CA, at an objective-lens focal length of 2.8 mm. The DNAs will be faint, although they are easily found once one is accustomed to looking for them.

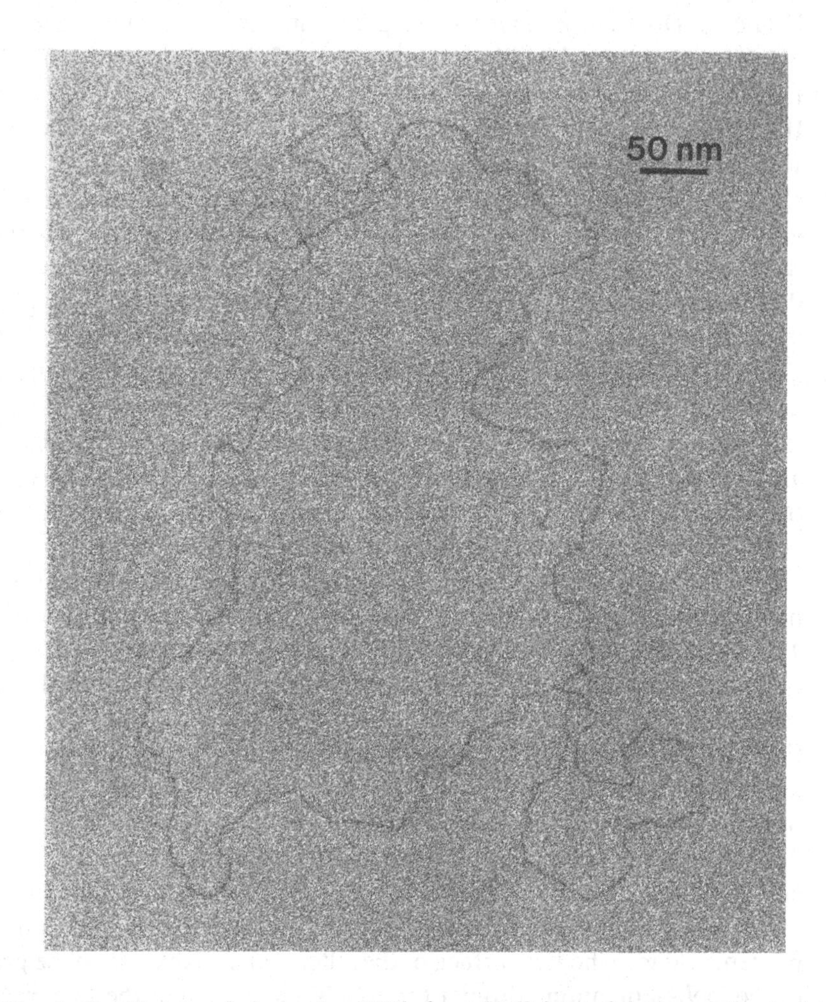

Fig. 1. Double-stranded circular DNA from PM2 bacterial virus, adsorbed to bacitracin and positively stained with uranyl acetate. Note the uniform adsorption, without stretching or backfolding, along the entire length of this DNA containing approx 7900 bp. Scale bar = 50 nm.

Figure 1 shows an example of the results obtained with this method. In the figure is shown a double-stranded circular DNA from PM2 virus, nicked to relax the superhelical twists. The contour of the DNA helix can be followed along its entire length, and localized short-range curvatures are visible. The uniform, undistorted adsorption of the DNA, the even background, and the exceptionally smooth trace formed by the uranyl acetate can all be attributed at least in part to the wetting

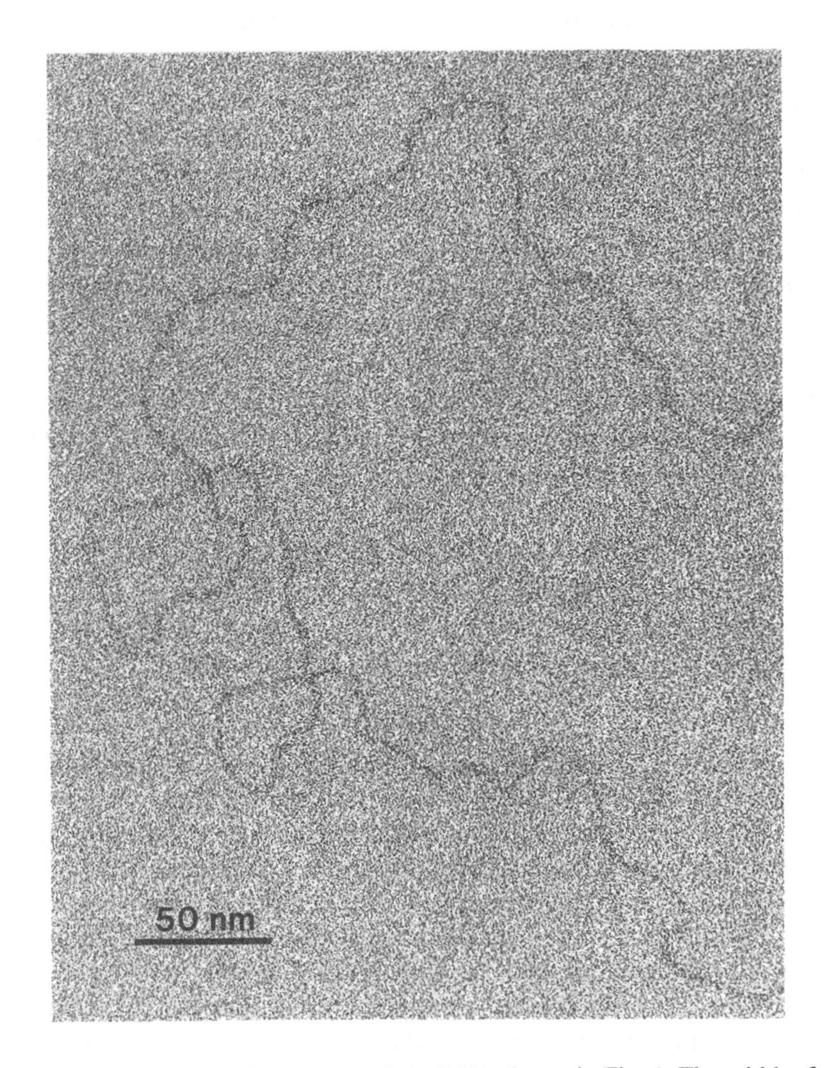

Fig. 2. Enlargement of a segment of the DNA shown in Fig. 1. The width of the continuous line of stain is essentially the same as the actual 2 nm diameter of the DNA helix, so that the DNA contours are visualized in true proportion to the helix width. Scale bar = 50 nm, equivalent to 147 bp of B-form DNA.

properties of the bacitracin *(8)*. Figure 2 shows an enlargement of a segment of the same DNA. The 50-nm scale bar corresponds to 147 bp of B-form DNA; localized curved regions involving as few as 40 bp can be clearly distinguished. The minimum measurable contour length is about 10–20 bp, corresponding to one or two helical turns. The width of the trace of uranyl acetate stain representing the DNA double helix is not more than the actual helix diameter of 2 nm. There is good contrast between the trace and the background, enabling one to follow the contours of the DNA readily.

For comparison, Fig. 3 shows a segment of a double-stranded DNA mounted in my laboratory by the Inman version *(10)* of the commonly used cytochrome *c* technique. The magnification is similar to that of Fig. 1. The DNA in Fig. 3 is greatly thickened by cytochrome *c* protein that is aggregated around it and it is thickened to a lesser extent by the metal deposits produced during shadowcasting. As a result, localized curvatures in the DNA helix are hidden, and it is impossible to determine accurately the lengths of shorter segments of a DNA that might be involved, for example, in loop formation (*see* Section 1.). Proteins bound specifically to the DNA will also be hidden by the cytochrome *c* coating.

Figure 4 shows the distribution of the lengths of 26 molecules of circular, double-stranded PM2-DNA adsorbed to bacitracin and positively stained with uranyl acetate. A unique length is defined, as expected; the standard deviation of the measured lengths is ±3%.

4. Notes

1. In examining our preparations of DNA adsorbed to bacitracin-coated carbons, we find that a layer of negative stain tends to persist around the edges of each square that is framed by the bars of the copper mesh supporting grid. This appears to be owing to a meniscus effect; the carbon film overlying the copper grid sags slightly into the unsupported area within each square, and residual stain-containing liquid remains trapped next to the raised edges. The DNAs in these darkly stained areas tend to be more highly folded, having probably been released from adsorption to the bacitracin during drying in the presence of the concentrated uranyl salts. The lightly stained areas in the center of each square are best; the DNAs there are generally free of localized folding or stretching, having the appearance of the DNA in Fig. 1. The stain intensity can be adjusted, if necessary, by doubling or halving the concentration of the uranyl acetate solution.
2. The grids can be shadowed, if desired, after examining and even making micrographs from the stained preparations. Shadowing is carried out with the grids at an angle of 20° to a source of evaporating tungsten, and the tungsten is deposited to a measured thickness of 1.5 nm. *(See* Chapter 8 for a description of shadowcasting methods.)

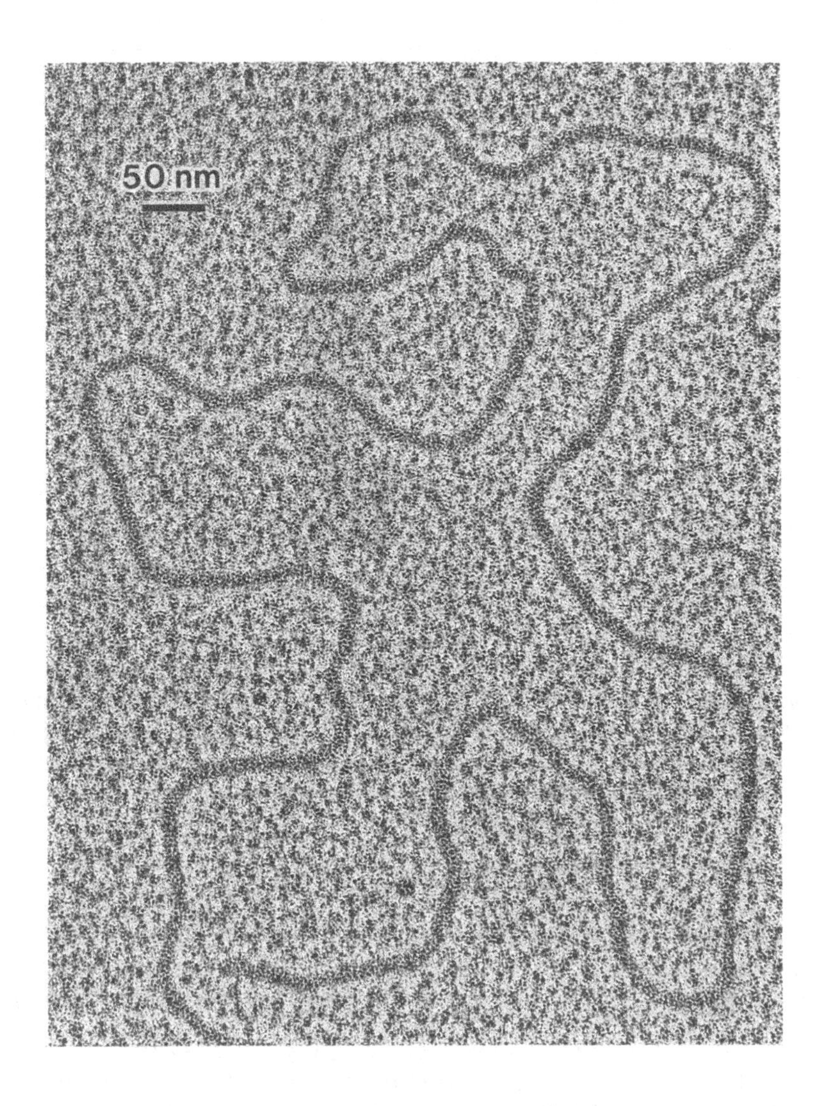

Fig. 3. Segment of a double-stranded linear DNA from adenovirus type 2, mounted by the Inman version *(10)* of the cytochrome *c* technique and rotary-shadowed with platinum-palladium. Note the exaggerated thickness of the DNA and the lack of detail of localized DNA contours, in comparison with Figs. 1 and 2. Scale bar = 50 nm.

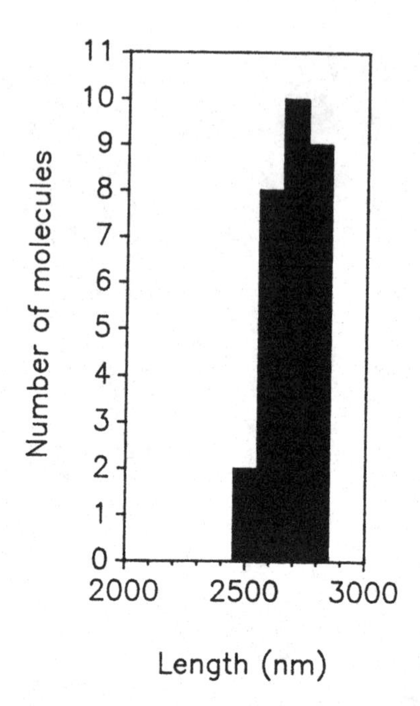

Fig. 4. Measured lengths of PM2-DNA molecules, mounted on a bacitracin film as in Figs. 1 and 2. Length measurements were performed on images projected from negatives, using a Numonics Model 1240 graphics calculator. The calculated mean length is 2680 nm ± 3%.

Acknowledgments

The author wishes to acknowledge excellent assistance with the length measurements by Y. T. Yang. This work was supported by National Institutes of Health Research Grant 5-R01-GM34293-03 (to C. W. G.), by NIH Biomedical Research Support Grant 2S07-RR07133-21, by NIH Small Instrumentation Grant 1-S15-NS25421-01, and by National Science Foundation Instrumentation Grant PCM-8116109.

References

1. Kleinschmidt, A. K. (1968) Monolayer techniques in electron microscopy of nucleic acid molecules, in *Methods in Enzymology* (Kaplan, N. O. and Colowick, S. P., eds.), vol. 12: *Nucleic Acids, part B* (Grossman, L. and Moldave, K., eds.), Academic, New York, pp. 361–377.
2. Lang, D. and Mitani, M. (1970) Simplified quantitative electron microscopy of biopolymers. *Biopolymers* **9,** 373–379.
3. Davis, R. W., Simon, M., and Davidson, N. (1971) Electron microscope heteroduplex methods for mapping regions of base sequence homology in nucleic

acids, in *Methods in Enzymology* (Kaplan, N. O. and Colowick, S. P., eds.), vol. 21: *Nucleic Acids, part D* (Grossman, L. and Moldave, K., eds.), Academic, New York, pp. 361–377.
4. Vollenweider, H. J., Sogo, J. M., and Koller, Th. (1975) A routine method for protein-free spreading of double- and single-stranded nucleic acid molecules. *Proc. Natl. Acad. Sci. USA* **72,** 83–87.
5. Griffith, J., Huberman, J. A., and Kornberg, A. (1971) Electron microscopy of DNA polymerase bound to DNA. *J. Mol. Biol.* **55,** 209–214.
6. Williams, R. C. (1977) Use of polylysine for adsorption of nucleic acids and enzymes to electron microscope specimen films. *Proc. Natl. Acad. Sci. USA* **74,** 2311–2315.
7. Gray, D. M., Ratliff, R. L., Antao, V. P., and Gray, C. W. (1988) CD spectroscopy of acid-induced structures of polydeoxyribonucleotides: importance of $C{\bullet}C^+$ base pairs, in *Structure and Expression, vol. 2: DNA and Its Drug Complexes* (Sarma, M. H. and Sarma, R. H., eds.), Adenine, New York, pp. 147–166.
8. Gregory, D. W. and Pirie, B. J. S. (1973) Wetting agents for biological electron microscopy. I. General considerations and negative staining. *J. Microscopy* **99,** 261–265.
9. Gregory, D. W. and Pirie, B. J. S. (1973) Wetting agents for biological electron microscopy. II. Shadowing. *J. Microscopy* **99,** 267–278.
10. Inman, R. B. (1974) Denaturation mapping of DNA, in *Methods in Enzymology* (Kaplan, N. O. and Colowick, S. P., eds.), vol. 29, *Nucleic Acids and Protein Synthesis, part E* (Grossman, L. and Moldave, K., eds.), Academic, New York, pp. 451–458.

CHAPTER 3

Biological Applications of Scanning Tunneling Microscopy

Phillip M. Williams, Mano S. Cheema, Martyn C. Davies, David E. Jackson, and Saul J. B. Tendler

1. Introduction

The scanning tunneling microscope (STM) is a new and exciting method of direct surface analysis. Following the microscope's first construction by Binnig and Rohrer in 1982 *(1,2)*—for which they won the 1986 Nobel Prize for Physics—the instrument has been extensively used to investigate the surface properties of many inorganic conducting materials. In recent years, the microscope has been utilized to investigate biological molecules deposited on suitable conducting surfaces, providing atomic resolution images of single molecules, with no conformational averaging as occurs for spectroscopic techniques associated with the study of bulk molecules. These studies show that the technique is a potentially valuable biophysical tool complementary to the other well established methods that are extensively reviewed in this volume. To date, high resolution images of biological systems, such as DNA *(3)*, globular macromolecules, such as vicilin *(4)*, and phospholipid membranes *(5)* have been obtained, with the body of scientific literature increasing rapidly with time. This chapter reports on the basis of the use of the technique for imaging biologicals, the equipment required, and how STM imaging is undertaken.

From: *Methods in Molecular Biology, Vol. 22: Microscopy, Optical Spectroscopy, and Macroscopic Techniques* Edited by: C. Jones, B. Mulloy, and A. H. Thomas

To illustrate the use of the technique on inorganic materials, Figs. 1 and 2 show unprocessed images of gold and graphite substrates, respectively. The quality of data obtained with biological samples is shown in Figs. 3–5, where we present directly obtained unprocessed images of an antigenic peptide, helical fibers of poly-benzyl-L-glutamate, and calf thymus DNA, respectively. The images were obtained in our laboratory on a VG Microtech (Uckfield, UK) STM 2000 operating in air, at ambient temperature and pressure.

The images obtained by STM depend on monitoring a small, but measurable current flowing between a tip and the sample surface when a voltage is applied. This is known as a tunneling current since, in quantum mechanical terms, the electron travels through a tunnel that connects the electron cloud of the tip with the electron cloud of the surface. Although most biological molecules are extremely poor conductors, it is normally found that they modify the tunneling barrier to a sufficient extent that they can, if relatively thin, be imaged directly. Thicker biologicals and nonconducting materials, such as thick polymer film insulation, may require the surface to be coated with a suitable conducting material, as discussed below. In the imaging process, the tip is scanned across the sample and the tunneling current monitored. In the constant current mode of instrument operation (the normal operation mode of the instrument), a feedback loop attempts to keep the tip at a constant height above the surface. If the distance between the tip and the surface decreases, the monitored current increases exponentially, and a feedback loop raises the tip to restore the current to its predetermined value and hence the original tip height above the surface. Conversely, if the distance between the tip and the surface increases, an exponential decrease in the monitored current results in the tip being lowered to restore the gap width. The tip therefore follows an undulating course above the sample as shown in Fig. 6.

An alternative to the constant current mode of STM imaging is the constant height mode of operation. In this mode, the tip is not allowed to move in the vertical direction; as the height between the tip and the surface varies, it produces fluctuations in the tunnel current. In general, this form of operation is only suitable for the imaging of near atomically flat surfaces. Tunnel current images can provide complementary data to the images obtained in the constant current mode. A

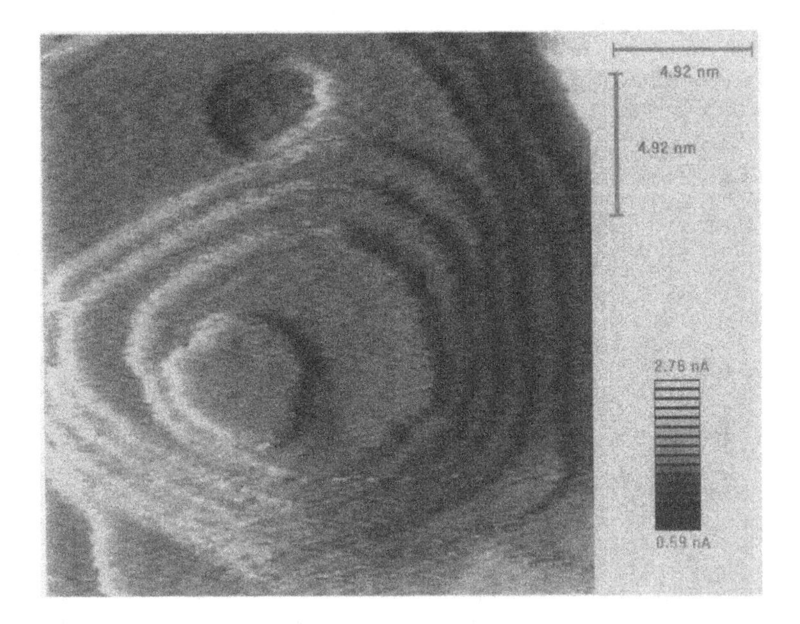

Fig. 1. STM image of gold surface (20 × 20 nm) showing atomic steps. (Constant current mode, current 1.0 nA, bias voltage 821 mV.)

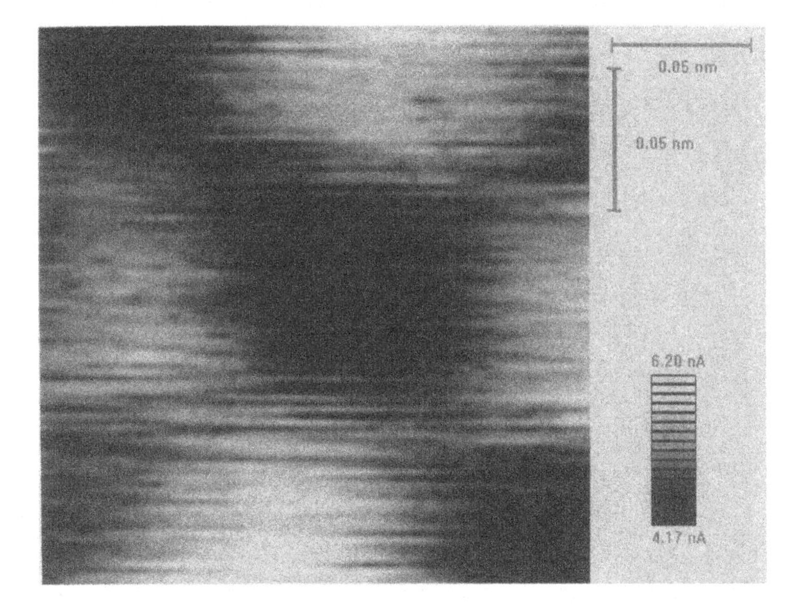

Fig. 2. Ultra-high-resolution STM image of HOPG (0.2 × 0.2 nm) showing four single carbon atoms. (Constant current mode, current 5.0 nA, bias –20 mV.)

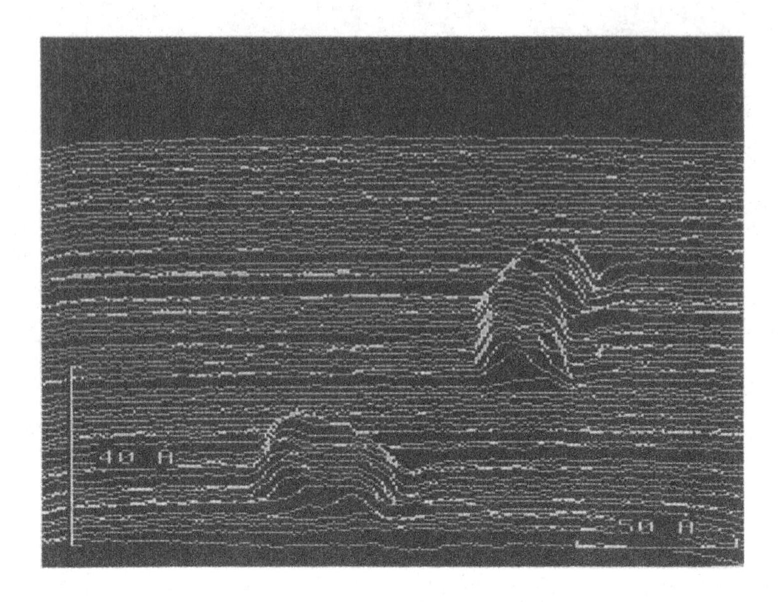

Fig. 3. Orthogonal view of two molecules of a 20 amino acid antigenic peptide in two distinct conformations.

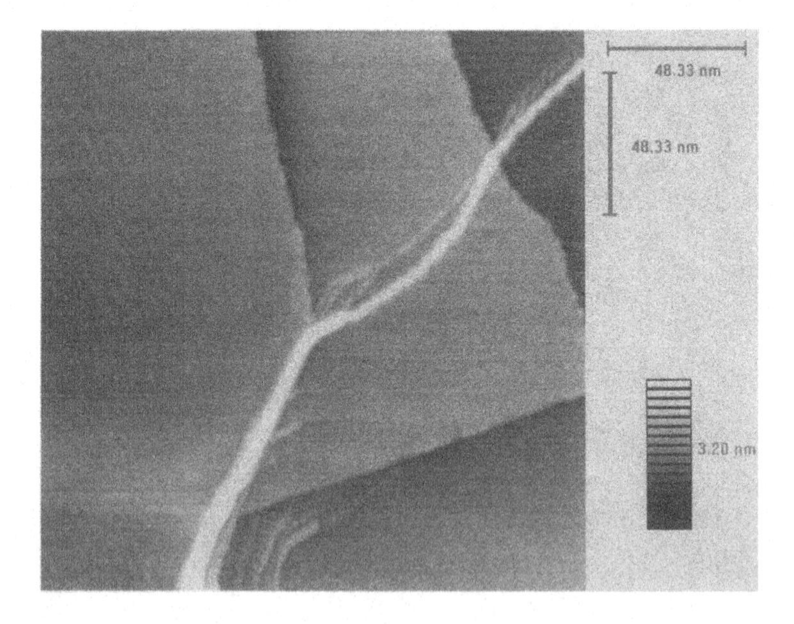

Fig. 4. Poly-benzyl-L-glutamate aggregates deposited across surface steps of HOPG. (193 × 193 nm, constant current mode, current 1.0 nA, bias voltage 100 mV.)

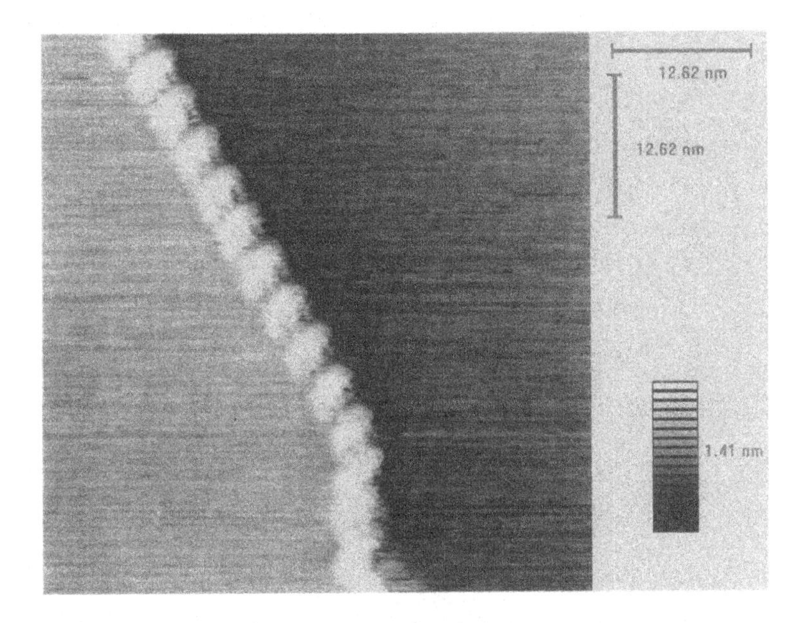

Fig. 5. Single strand of calf thymus DNA deposited along a surface step of HOPG. (50 × 50 nm, constant current mode, current 0.1 nA, bias voltage 500 mV.)

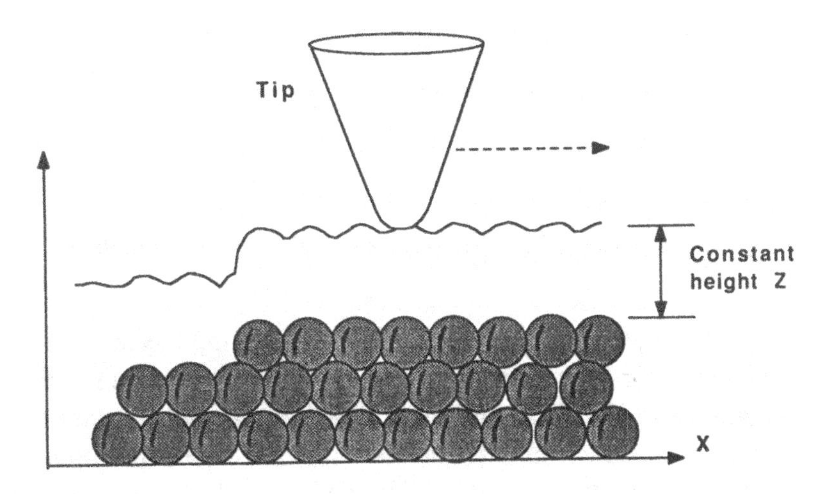

Fig. 6. Schematic representation of the probe tip scanning the surface in constant current mode.

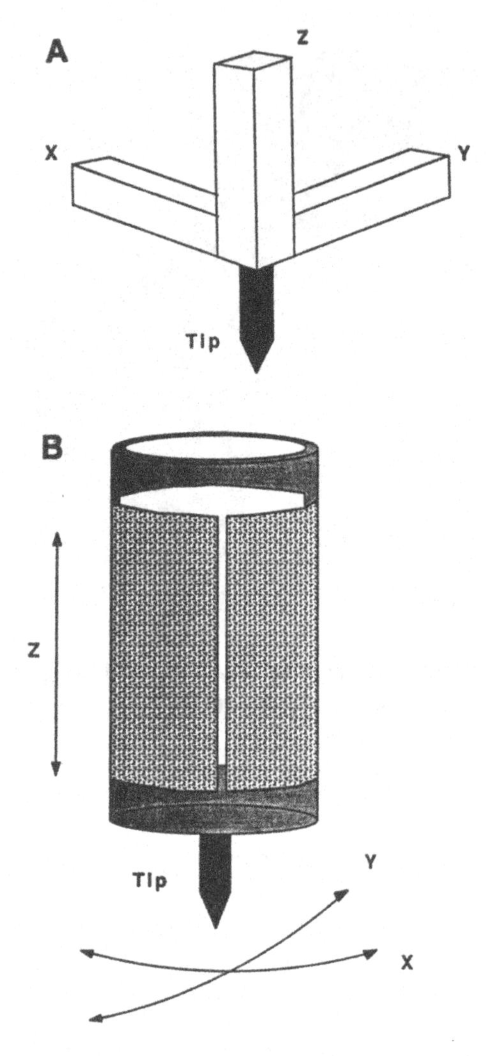

Fig. 7. Tripod (A) and tube (B) arrangement of piezo-electric transducers.

detailed investigation of the theory of STM tunneling has been undertaken by a number of groups *(6,7)*, and the subject has been reviewed comprehensively *(8)*. The tip is positioned by high-quality piezo-electric transducers, which expand on the order of angstroms upon the application of a suitable voltage. The early STM instruments used three independent lead titanate-lead zirconate piezo-electric transducers (x,y,z) arranged in a tripod arrangement as shown in Fig. 7A, and

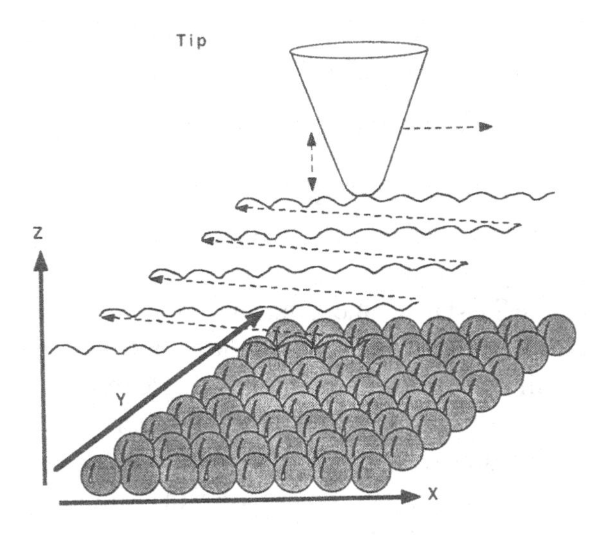

Fig. 8. Tip rastering across the surface.

scanning was achieved by the application of suitable voltages to the x and y piezo-electric transducer. The height between the tip and the surface was maintained using the z piezo. In modern STM instruments, the tripod drive has been replaced by a cylindrical piezo-electric tube (Fig. 7B), which has electrodes mounted on it. These electrodes control the ±x and ±y movement; a single electrode in the tube controls ±z motion.

In normal STM imaging (depending on the instrument), a voltage of the order of 0.002–2 V is maintained between the tip and the sample, and the observed current is in the nanoampere range. Imaging is carried out by rastering the tip across the sample, as shown in Fig. 8. In the constant current mode of operation, plots of tip rise (i.e., voltage applied to the z piezo-electric transducer—z-voltage) against scan distance traveled provide a topographical map of the surface. The displayed image is a representation of z(x,y), for example, as a 256 × 256 pixel (dot) computer image with the gray scale intensity or false color of each pixel representing "height" above the substrate. In most cases, the data are displayed in plan view (Figs. 1,2,4,5) or as an orthogonal projection (Fig. 3).

In optical and electron microscopy, the resolution obtained is limited by the wavelength of electromagnetic radiation used to image the sample. This obviously is not a limitation for the STM where the use of nanotechnology provides the greatest power of resolution of any microscopy technique. The STM has an optimum lateral resolution of 1 Å and a vertical resolution of <0.1 Å.

2. Materials and Methods

2.1. Sample Purity

For accurate image interpretation, the sample and all materials associated with the imaging process should be of the highest purity available. However, the STM instrument may be employed to image different molecules and their differing conformations within a heterogeneous mixed sample. Since one normally images single molecules, the quantity of the sample required is correspondingly very low, normally on the milli- to nanogram scale. In order to avoid particulate contamination, all samples should, ideally, be prepared in a suitable laminar flow cabinet using aseptic techniques and with standard surface analytical procedures undertaken throughout *(9)*. The detailed choice and specifications of the substrates and tips are discussed at length below.

2.2. The Choice of Conducting Substrate

In general, STM image data of a biomolecule are recorded after the sample has been deposited onto a suitable molecularly smooth conducting substrate. In the majority of cases, highly oriented pyrolytic graphite (HOPG) has been employed, but care must be taken to distinguish substrate artifacts that may resemble biomolecular structure. HOPG is readily available commercially and, when used as 5 × 5 mm sheet 2-mm thick, is easily cleaved to reveal a fresh, near atomically smooth surface. The cleaving of the surface plane is achieved simply, by applying a piece of adhesive tape to the surface and removing it carefully. Other materials that have been employed for STM analysis of adsorbed molecules include molybdenum sulfide single crystals, gold, and platinum/carbon films on freshly cleaved mica. The latter substrate may offer some advantage because of its reputedly stronger adsorption molecules, reducing the chance of lateral sample migration through interaction with the scanning tip. This may be a problem with weakly adsorbed materials.

2.3. Sample Preparation and Deposition

Samples can be presented to the conducting substrate by a variety of means. In the majority of cases, however, the sample is prepared simply by depositing a few microliters of a dilute solution onto the appropriate substrate. A range of organic solvents and water may be employed. Excipients may be included in the solution to facilitate sample imaging. In studies on calf thymus DNA *(3)*, for example, spermidine hydrochloride or hexamine cobalt (III) has been added to promote lateral association by charge neutralization. The authors suggested *(3)* that image data obtained from molecular bundles rather than individual molecules appeared to give the most regular and reproducible images possibly through the mutual stabilization of neighboring molecules under the dehydrated conditions.

The influence of substrate on the molecular ordering of adsorbed aggregates must also be considered. The orientation of 4-*n*-octyl-4-cyanobiphenyl molecules in smectic liquid crystals on HOPG and molybdenum sulfide has been shown to be markedly different because of preferential positioning of the two surfaces *(10)*. In most cases, STM image data have been recorded in the absence of the solvent, which has been removed either by slow freeze-drying as employed in electron microscopy or, more commonly, by allowing evaporation to occur under ambient conditions within a laminar flow environment. The range of biological materials imaged in this manner in air include DNA *(3)*, polypeptides *(11)*, proteins *(4)*, and polysaccharides *(12)*. Although these studies provide detailed information about the conformation of biological molecules deposited on inorganic materials, the ultimate aim of many of the biophysical studies is to relate the conformations observed with that found in the aqueous environment. Successful attempts have now been made to image DNA in water *(13)*. The DNA was retained at a submerged gold electrode by electrophoretic deposition. The DNA helices were visualized with sufficient resolution to identify the 3.4-nm helix of the B-form of DNA.

To overcome the poor conductivity of larger structures, such as globular macromolecules, aggregates, and cellular materials, and hence to improve the image resolution, the application of metal coatings to the sample surface has been advocated by a number of researchers. This has been achieved by coating with 1 nm layers of platinum-iridium-carbon alloy using commercially available rotary shadowing equipment. This procedure may be particularly advantageous for imaging large molecular aggregates or whole-cell structures, such as viruses, although to date STM images obtained in this way are, in general, not much more structurally informative than conventional electron micrographs.

The analysis of dynamic interfaces, such as membranes, has been undertaken by simple deposition onto HOPG, but higher resolution data have been obtained using the more satisfactory method of freeze-fracture replication *(14)*. The process involves the rapid freezing of the system, followed by fracture *in vacuo* on a cryogenic stage, and subsequent coating with an evaporated layer of platinum or carbon. A guide to the subtlety of this technique is published elsewhere *(5)*.

2.4. The Choice of Tip

The condition and nature of the tip employed in the STM experiment will markedly influence the quality of the image data obtained. Indeed, the radius of curvature at the tip apex actually determines the lateral resolution of the STM. Platinum-iridium (Pt-Ir), gold, and tungsten are routinely employed for tip construction. Fortunately, it is relatively simple to prepare a tip to the required atomic resolution at its apex. Pt-Ir tips may be simply cut or mechanically ground, whereas the tungsten tip may be formed by etching in a potassium hydroxide solution. On the practical level, a tip that appears sharp under a good-quality optical microscope will in general give atomic resolution.

Tip cleanliness during initial preparation is of utmost importance. Once the tip has been mounted using aseptic-style handling, performance can be enhanced by further treatment. A number of methods have been used to achieve *in situ* cleaning or reconditioning of the STM tip, as detailed by Hansma and Tersoff *(8)*.

2.5. General Operating Instructions

The following is a general guide to how one obtains high-resolution STM images of biological systems deposited on a suitable conducting substrate. Although the operations of most commercially available instruments are similar, these instructions are obviously instrument-dependent and are based on our own STM, a VG Microtech STM 2000, shown in Figs. 9 and 10.

1. The conducting substrate with the deposited sample is mechanically raised to within a few nanometers of the tip. This maneuver is normally facilitated by the use of the instrument's optical microscope. The lateral position of the sample is adjusted using micrometers that move the stub in both the x and the y directions, allowing the correct portion of the substrate to be imaged.

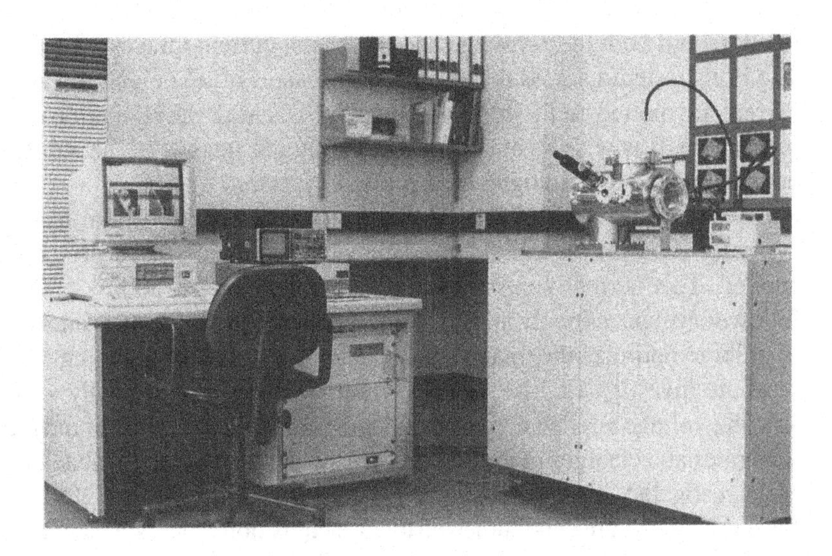

Fig. 9. VG Microtech STM 2000 unit comprising head and chamber on antivibration cabinet, acquisition unit, and data management work station.

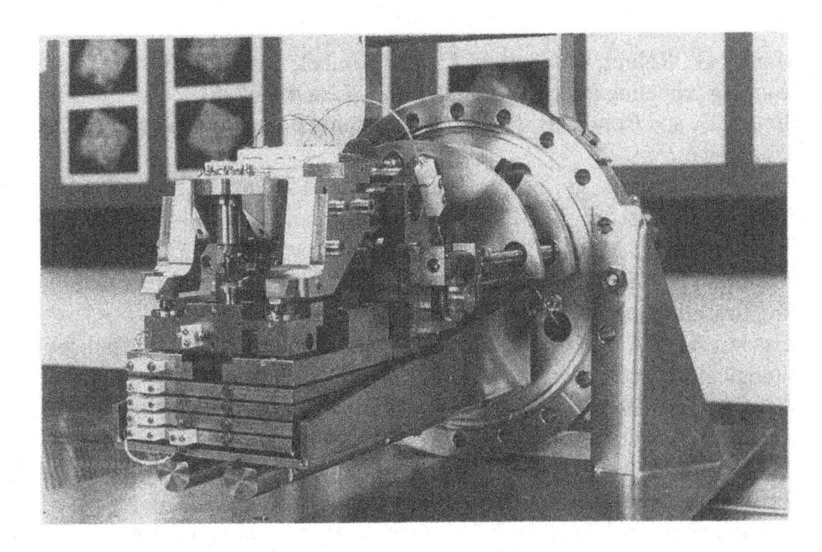

Fig. 10. STM head and stage (VG Microtech STM 2000).

2. To start the tunneling process, the stub is slowly raised using a stepper motor under computer control until a suitable tunnel current flows (e.g., 1 nA). The STM is now ready to obtain images. In our laboratory, we routinely record both the z-voltage and the tunnel current images.
3. The tip is rastered across the sample, and images of tip height and tunnel current are plotted on the computer display. To check the resolution of the instrument, it is wise to obtain atomic resolution images of the substrate before imaging the biological molecules of interest.

   Because of the relatively low concentrations of biological molecules normally used on the conducting substrate, a simple search protocol is applied. This normally initially entails large area scans (5000 × 5000 Å) followed by successively smaller scans centered about areas of interest.
4. In order to optimize the images, the effects of a variety of scanning conditions are investigated. The tunnel current maintained is normally varied, as is the voltage applied across the tip and sample. The effects of different scan rates and changes in the electrical response time of the feedback loop are investigated.
5. Images of interest may be subjected to postacquisition processing and analysis using a variety of techniques, including Fourier transformation. The data may also be transferred into a computational chemistry/molecular modeling environment, where the images obtained can be related to theoretical and experimental data obtained from such techniques as X-ray crystallography and NMR spectroscopy.

References

1. Binnig, G., Rohrer, H., Gerber, C., and Weibel, E. (1982) Surface studies by scanning tunneling microscopy. *Phys. Rev. Lett.* **49,** 57–61.
2. Binnig, G. and Rohrer, H. (1985) The scanning tunneling microscope. *Sci. Am.* **253,** 50–56.
3. Arscott, P. G., Lee, G., Bloomfield, V. A., and Evans, D. F. (1989) Scanning tunneling microscopy of Z-DNA. *Nature (London)* **119,** 484–486.
4. Welland, M. E., Miles, M. J., Lambert, N., Morris, V. J., Coombs, J. H., and Pethica, J. B. (1989) Structure of the globular protein Vicilin revealed by scanning tunneling microscopy. *Int. J. Biol. Macromol.* **11,** 29–32.
5. Zasadzinski, J. A. N. (1989) Scanning tunneling microscopy with applications to biological surfaces. *Biotechniques* **7,** 174–187.
6. Tersoff, J. and Hamann, D. R. (1983) Theory and applications for the scanning tunneling microscope. *Phys. Rev. Lett.* **50,** 1998–2001.
7. Baratoff, A. (1984) Theory of scanning tunneling microscopy—methods and applications. *Physica (Utrecht)* **127B,** 143–150.
8. Hansma, P. K. and Tersoff, J. (1987) Scanning tunneling microscopy. *J. Appl. Phys.* **61,** R1–458.
9. Mittal, K. (ed.) (1979) *Surface Contamination*. Plenum, New York.

10. Hara, M., Iwakabe, Y., Tochigi, K., Sasabe, H., Garito, A. F., and Yamada, A. (1990) Anchoring structure of smectic liquid-crystal layers on molybdenum sulphide observed by scanning tunneling microscopy. *Nature (London)* **344,** 228–230.
11. McMaster, T. J., Carr, H. J., Miles, M. J., Cairns, P., and Morris, V. J. (1990) Scanning tunneling microscopy of poly-benzyl-L-glutamate. *J. Vac. Sci. Technol.* **A8,** 672–677.
12. Miles, M. J., McMaster, T., Carr, H. J., Tatham, A. S., Shewry, P. R., Field, J. M., Belton, P. S., Jeens, B., Hanley, M., Whittam, P., Cairns, P., Morris, V. J., and Lambert, N. (1990) Scanning tunneling microscopy of biomolecules. *J. Vac. Sci. Technol.* **A8,** 698–702.
13. Linsay, S. M., Thundat, T., Nagahara, L., Knipping, U., and Rill, R.L. (1989) Images of the DNA double helix in water. *Science* **244,** 1063,1064.
14. Zasadzinski, J. A. N., Schneir, J., Gurley, J., Elings, V., and Hansma, P. K. (1988) Scanning tunneling microscopy of freeze-fractured replicas of biomembranes. *Science* **239,** 1013,1014.

PART II

SCATTERING AND SEDIMENTATION

CHAPTER 4

High-Flux X-Ray and Neutron Solution Scattering

Stephen J. Perkins

1. Introduction

Solution scattering is a diffraction technique that is used to study the overall structure of biological macromolecules in the solution state *(1–3)*. Although X-rays are diffracted by electrons and neutrons are diffracted by nuclei, the physical principles are the same. Scattering views structures in random orientations to a nominal structural resolution of about 2–4 nm in a Q* range ($Q = 4\pi \sin\theta/\lambda$; 2θ = scattering angle; λ = wavelength) between about 0.05 and 3 nm^{-1} (Fig. 1). In comparison, the use of diffraction to study single crystals of macromolecules will lead to electron or nuclear density maps at atomic resolution (0.15 nm) as the result of crystalline order. Analyses of the scattering curve $I(Q)$ measured over a range of Q lead to the mol wt and the degree of oligomerization, the overall radius of gyration R_G (and in certain cases, those of the cross-section and the thickness), and the maximum dimension of the macromolecule. Scattering can be used to monitor conformational changes. The scattering analyses can be quantitatively compared with other physical data in order to check and refine the results. These other methods include electron microscopy (Chapters 1 and 2), determinations of sedimentation or diffusion coeffi-

*Q is a measure of the scattering angle and is a convenient algebraic abbreviation. Note that $Q = 2\pi/d$, where d is the diffraction spacing specified in Bragg's Law of Diffraction: $\lambda = 2d \sin\theta$.

From: *Methods in Molecular Biology, Vol. 22: Microscopy, Optical Spectroscopy, and Macroscopic Techniques* Edited by: C. Jones, B. Mulloy, and A. H. Thomas

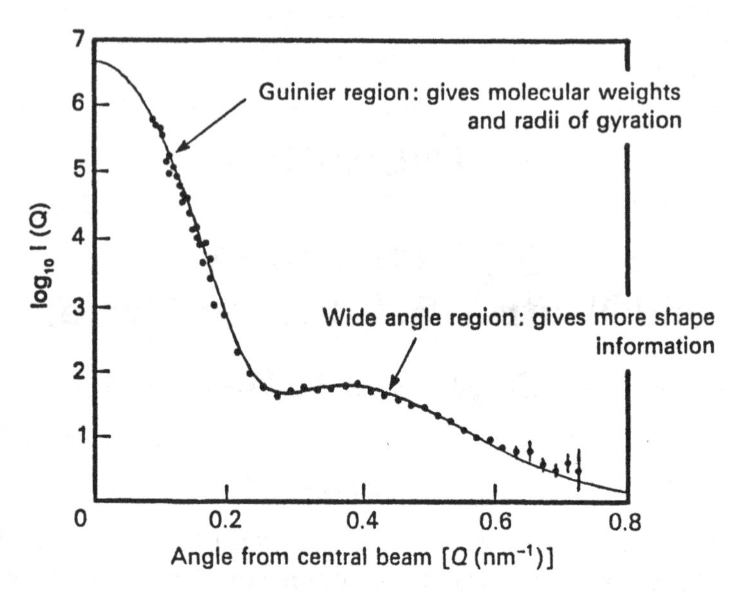

Fig. 1. General features of a solution scattering curve I(Q) measured over a Q range *(2)*. The neutron scattering curve of complement component C1q in 100% D_2O buffers is analyzed in two regions, that at low Q, which gives the Guinier plot from which the overall radius of gyration R_G and the forward scattered intensity I(0) values are calculated, and that at larger Q, from which more structural information is obtained. At low Q, the scattering curve is truncated for reason of the beamstop. The scattering curve was measured using two sample-detector distances of 2.7 and 10.7 m on instrument D11 at the ILL Grenoble; the shorter distance defines the maximum Q measured.

cients (Chapters 5 and 6), crystallography (*Methods in Molecular Biology: Crystallographic Methods and Techniques* [in press]), and molecular graphics modeling. The most important feature of solution scattering is that it offers a multiparameter characterization of the overall structure under physiological conditions. In this respect, it is superior to electron microscopy and hydrodynamic methods. Ideally, solution scattering is applied to the study of a well-characterized macromolecule of known sequence in order to determine one or two parameters that were previously unknown. For example, using the structure of a single domain in a multidomain protein as a constraint, the overall arrangement of the domains in this protein under physiological conditions can be analyzed.

There are several advantages of high-flux sources for solution scattering. The high flux means that X-ray scattering cameras can utilize ideal point collimation geometries (i.e., based on pin-hole optics) and a highly monochromatized beam. These avoid scattering curve distortions, and signal-to-noise ratios are dramatically improved. These make possible the use of time-resolved scattering to follow the rate of conformational changes or oligomerization/dissociation processes; and the study within one experimental session (1–2 d) of a large number of samples. Also, the need for nonphysiological high sample concentrations is minimized, and macromolecules of low solubility can be studied.

Analogous advantages hold for neutron scattering. The high flux additionally makes possible the use of contrast variation experiments in mixtures of H_2O and D_2O to investigate the internal structure of the macromolecule. The interiors of biological macromolecules are generally inhomogeneous with respect to their scattering properties. Thus lipids, proteins, carbohydrates, and nucleic acids are each characterized by their own distinct scattering density. H_2O and D_2O have very different scattering densities. Variations of the ratio of these two solvents in neutron scattering experiments will reveal the structure of these components within the macromolecule. X-ray data can be used as a substitute for neutron data in 0% D_2O. This is exemplified for low-density lipoprotein (LDL) in Fig. 2 *(4)*. LDL is almost spherical in its shape and is formed from apolipoprotein B (mol wt 540,000) in association with lipids to result in a total mol wt close to 2,300,000. The high positive solute–solvent contrast found with X-ray scattering emphasizes the protein component of LDL, whereas the high negative solute–solvent contrast in D_2O by neutron scattering emphasizes the lipid content. The resulting scattering curves are markedly different in the positions of the maxima and minima, and can to a first approximation be analyzed in terms of a simple two-density shell model.

1.1. Outline of the Method of Solution Scattering

Instrumental requirements are based on the irradiation of a solution of path thickness 1–2 mm with a collimated, monochromatized beam of X-rays or neutrons, and recording the scattering (or diffraction) pattern with a one- or two-dimensional detector linked to a minicomputer. Scattering cameras are provided and maintained as a multiuser facility by the institute providing the high-flux X-ray or neutron beams.

A synchrotron X-ray camera (such as at Stations 2.1 or 8.2 at SRS, Daresbury; *5,6*) takes a "white" beam of X-rays, which is emitted tangentially by the electrons circulating at relativistic speeds in the storage ring of the synchrotron (Fig. 3). The electron beam lifetime is

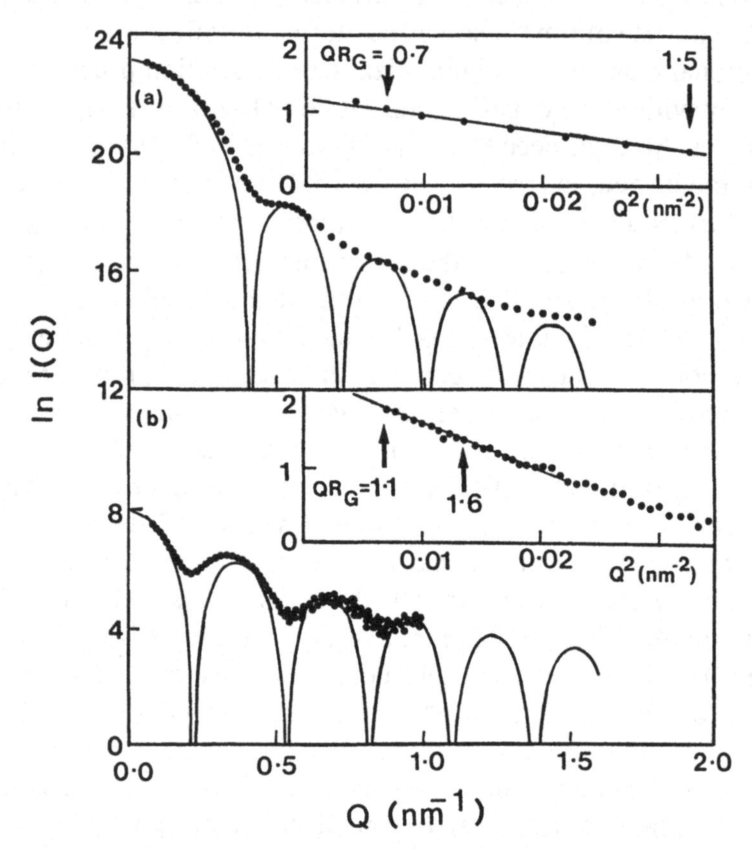

Fig. 2. Representative neutron (**a**) and synchrotron X-ray (**b**) scattering curves for native LDL in physiological buffers at 20°C. The Guinier plots to determine R_G and I(0) from the experimental curves are shown in the insets. The neutron Guinier fits were made in the Q range of 0.08–0.18 nm^{-1}, and the X-ray Guinier fits in the range 0.08–0.12 nm^{-1}. The results of a theoretical two-density shell modeling of the neutron and X-ray scattering curves for native LDL are shown as the continuous line. (**a**) The neutron model is defined by 11.0 nm (outer radius) and 10.5 nm (inner radius) with a ratio of scattering densities (outer:inner) of –1.0:–1.84 that is predicted from the calculated densities of glycoprotein and lipid. This is compared with the experimental scattering curve measured at 15 mg/mL. (**b**) The X-ray model is defined by 11.8 nm (outer radius) and 10.8 nm (inner radius) with a ratio of scattering densities (outer:inner) of 1.0:–0.13. This is compared with the experimental scattering curve measured at 10 mg/mL. That slightly larger dimensions are found with X-ray scattering is attributed to the hydration of LDL, which is detectable by X-ray scattering, but not in neutron scattering.

of the order of 12–24 h. The X-ray beam is horizontally focused and monochromated to a value close to 0.15 nm in wavelength, typically by a Ge or Si perfect single crystal, then vertically focused by a curved mirror, and collimated by slits *(6)*. Samples (1 mm path length; surface area 2 × 8 mm; total vol 25 µL) are held in brass water-cooled cells with 10–20 µm thick mica windows or in quartz capillaries. If the samples react with brass (a Cu-Zn alloy), the brass holder can be gold-plated; alternatively, perspex cells can be used. The sample holder is aligned in the beam by the use of "green paper," which turns red when exposed to X-rays. Sample-detector distances (0.5–5.0 m) depend on the desired Q range. Position-sensitive detectors are based either on a linear detector, which monitors the scattered intensities in one dimension, or on a quadrant detector, which monitors the intensities scattered in a two-dimensional angular sector (70°) of a circle with the nominal position of the main beam located at the center of the circle. The latter type leads to much improved counting statistics at large scattering angles, and a larger Q range is available, compared to the linear detector used for the data in Fig. 2(b). Beam exposures are monitored by the use of an ion chamber positioned after the sample; this method takes both the sample transmission and the flux into account. The detector is interfaced with a VAX minicomputer system for data storage and on-line processing to assess the experimental data as it is being recorded. The camera is inside a radiation-shielded hutch, protected by safety interlocks to avoid accidental lethal exposures to the users.

A neutron camera at a high-flux reactor (such as D11 or D17 at ILL, Grenoble; *7,8*) receives neutrons in a beam guide from the reactor after they have been moderated by a cold source (liquid D_2 at 25 K) to adjust the profile of neutron wavelengths to a maximum of 0.1–1 nm (Fig. 4). Beam monochromatization is achieved by use of a velocity selector based on a rotating drum with a helical slit in it, which allows through only the neutrons of desired speed (i.e., the wavelength). Beam collimation is by use of long segments (up to 40 m) of straight beam guides. The beam size is defined by a diaphragm. Alignment of the beamstop is performed by the use of a 0.5 mm thick Teflon™ strip, which is a strong isotropic scatterer. Up to 12 samples (1 or 2 mm path lengths; surface area 7 × 10 mm^2; total vol 150 or 300 µL) in commercially available standard rectangular quartz cells are loaded onto an auto-

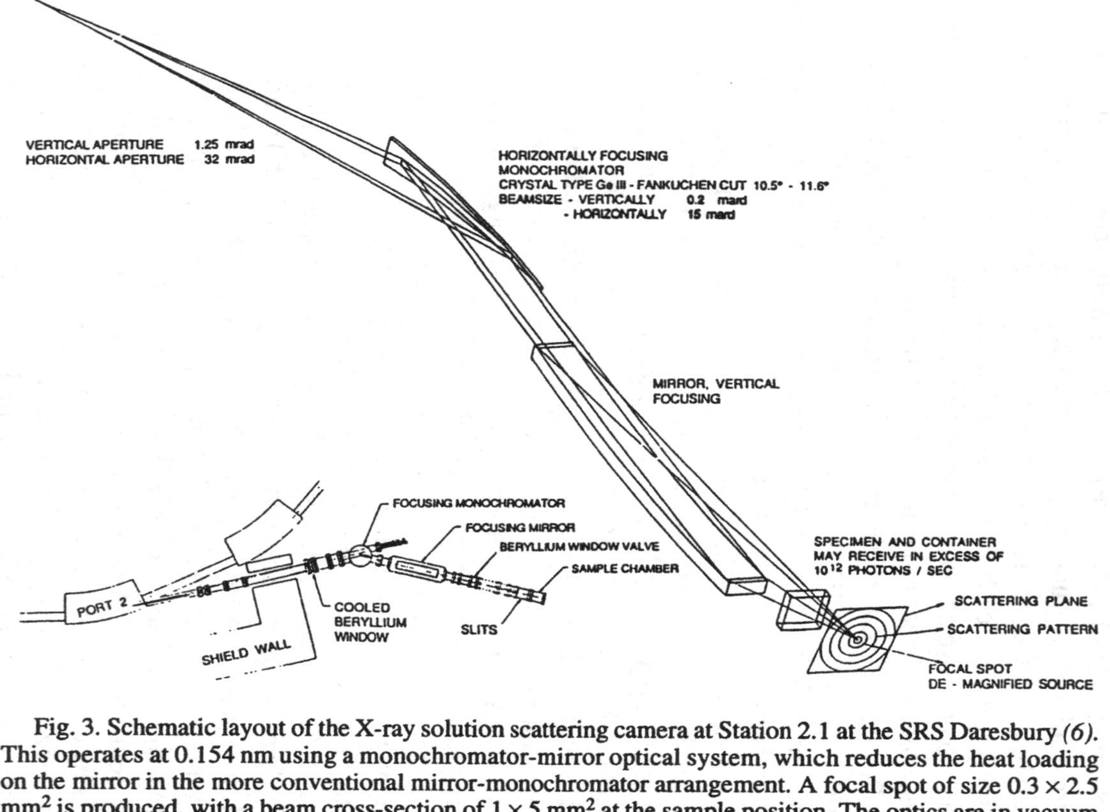

Fig. 3. Schematic layout of the X-ray solution scattering camera at Station 2.1 at the SRS Daresbury *(6)*. This operates at 0.154 nm using a monochromator-mirror optical system, which reduces the heat loading on the mirror in the more conventional mirror-monochromator arrangement. A focal spot of size 0.3×2.5 mm^2 is produced, with a beam cross-section of 1×5 mm^2 at the sample position. The optics are in vacuum and built on a vibration-isolation system. Between the sample and the detector (not shown) are sections of vacuum tubing of length between 0.5 and 5 m mounted on an optical bench. The scattering pattern is measured with either a linear, quadrant, or area detector that is interfaced to a minicomputer. Inset at the lower left is an overall view showing how the X-ray beam is taken from the synchrotron storage ring.

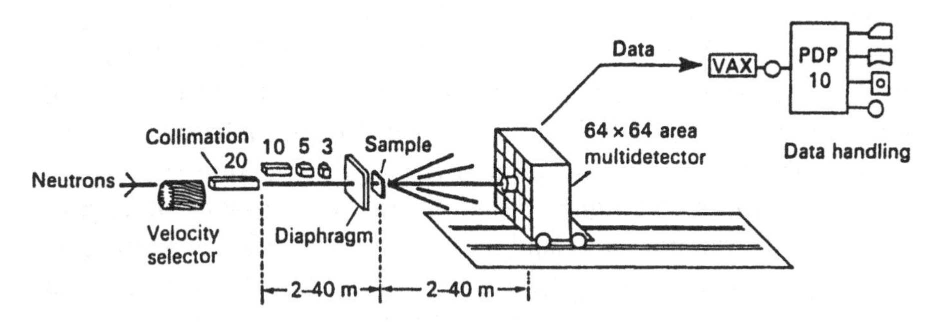

Fig. 4. Schematic view of the neutron solution scattering camera D11 at the ILL Grenoble *(3,7)*. A helical velocity selector is used for monochromatization. Movable neutron guides give collimation distances of 2, 5, 10, 20, and 40 m. The beam size at the sample is defined by a diaphragm. The 64 × 64 element BF_3 detector is housed within a 40-m evacuated tube and can be moved by remote control between 2 and 38 m sample-detector distances. The detector is interfaced with a minicomputer for data collection and storage.

matic sample changer, which is then programmed to run the samples in the neutron beam. Neutron exposures are monitored using a small sensor in the incident beam; data normalization therefore requires sample transmissions, which have to be measured separately. On D17, sample-detector distances are 0.8, 1.4, 2.8, or 3.5 m, with a choice of beam-detector axis angles from 0–90°, and neutron wavelengths are available between 0.8 and 2.0 nm. A typical configuration suitable for Guinier and wide-angle curves utilizes the 1.4 and 3.5 m distances with a wavelength of 1.1 nm (maximum neutron flux on D17). On D11, sample-detector distances are continuously variable between 2 and 38 m, and wavelengths range from 0.45–2.0 nm. A typical configuration is based on 2.5 and 10 m distances with a wavelength of 1.0 nm. It is helpful for subsequent analyses to keep the same wavelength between changes in the detector position. Two-dimensional multiwire BF_3 area detectors based on 10 × 10 mm^2 cells (64 × 64 in all) are used. Data acquisition, storage, and instrument control are performed by a VAX 11/730; on-line analyses of incoming data can be performed at the instrument to monitor progress. No radiation hutches are required; however, users should not approach the sample area when the beam is on.

A neutron camera at a spallation source (such as LO*Q* at ISIS, Rutherford Appleton Laboratories, Chilton, Oxon.; *9*) utilizes pulses of neutrons that are emitted from a uranium target after proton bombardment from a synchrotron. The main differences from a reactor camera are enumerated. The neutrons are moderated by a liquid D_2 cold source, so that each pulse is "white" with wavelengths in the range 0.2–1.0 nm. The collimation is 4.5 m in length up to the sample, whereas the sample-detector distance is 4.3 m. Monochromatization is achieved by time-of-flight techniques based on the total distance of 15.5 m between the uranium target and the BF_3 detector. After scattering by the sample, the neutrons in each pulse will reach the detector at different times depending on their wavelength. This has the advantage of efficiency in that all the neutrons in each pulse will be used. Transmissions at each wavelength are measured simultaneously using a semitransparent beamstop or on their own separately from scattering data collection. Wavelength selection choppers are used to control the neutron wavelength distribution within each pulse and prevent overlapping between consecutive pulses. An advantage of this camera is that the entire scattering curve between Q values of 0.06 and 2.2 nm^{-1} can be measured at one time with one sample-detector distance. A side benefit of this method is that neutrons of shorter wavelengths are the most abundant, and these contribute the most to the scattered curve at large Q where the intensities (and signal-to-noise ratios) are the lowest. However, the complexity of data processing means that the computational requirements to extract scattering curves are considerable.

1.2. Advantages and Disadvantages for Biology

The primary advantage of solution scattering in biology is that it is the only method that offers a multiparameter characterization of the gross structural features of macromolecules in a physiological environment. Even though electron microscopy has the ability to view structures directly, it has disadvantages in being prone to sample artifacts caused either by the need to work *in vacuo*, the use of stains, electron beam damage, magnification errors, and possible difficulties in interpretation. Since hydrodynamic methods give a single structural parameter based on the frictional coefficient, these report only on the degree of structural elongation. Although protein crystallography will yield structures at atomic resolution, crystallization buffers are usually unphysiological, and there will be no knowledge of how the gross structure behaves in solution. The latter is particularly important for multidomain structures with flexibility between the domains; often these cannot be crystallized intact.

The probability of a diffraction event when an X-ray photon or a neutron approaches an electron or nucleus is similar and very low at 10^{-25} or 10^{-23}, respectively. Compared to the use of solution scattering in chemistry and physics, the principle disadvantage of biological experiments lies in the signal-to-noise ratios since dilute solutions are employed. The use of high-flux sources minimizes constraints resulting from this limitation. Large-scale preparations involving about 50 mg of material and concentrations in the range of 5–10 mg/mL will be required for satisfactory investigations, although these concentrations are usually close to physiological values.

The application of the theory of solution scattering can be complex *(1)*. There is a possibility of overinterpretation of the scattering curves (i.e., reading too much into the analyses), or overlooking some detail during sample preparation or data collection. Collaborations or discussions with expert colleagues can be of value.

1.3. Comparison Between X-Rays and Neutrons

The two techniques are complementary in many respects. X-ray scattering has the following characteristics:

1. Most biological macromolecules are studied in high positive solute–solvent contrasts. This contrast corresponds to the situation in which the scattering density of the macromolecule is significantly higher than that of the solvent. This minimizes the systematic errors in the curve modeling of proteins that result if internal density fluctuations in the protein are neglected.
2. Good counting statistics are obtained in this contrast despite high background levels in the buffer curves, unlike neutron scattering in H_2O where the high incoherent scattering background of the buffer is a handicap.
3. Errors caused by wavelength polychromicity and beam divergence are not significant for synchrotron X-ray scattering; thus, Guinier and wide-angle analyses are not affected by systematic errors caused by the instrument geometry.
4. The hydrated dimensions of the macromolecule are studied, so the structure is larger by an additional depth of 0.36 nm at the surface to correspond to a monolayer of bound water molecules. For proteins, this corresponds to 0.3 g H_2O/g macromolecule.

Neutron scattering has the following characteristics:

1. Contrast variation in mixtures of H_2O and D_2O permits the analysis of hydrophobic and hydrophilic regions within proteins and glycoproteins, and the elucidation of the disposition of detergents or lipids in complexes with solublized membrane proteins (Fig. 2), or that of DNA or RNA in complexes with proteins. Deuteration of components in a multicomponent system extends these methods.
2. No radiation damage effects are encountered. This can be a severe problem with synchrotron X-rays. The neutron samples can normally be recovered for other studies.
3. The dry dimensions of the macromolecule are studied and correspond to the macromolecular structure observed by protein crystallography.
4. Absolute mol-wt calculations are obtained from neutron data in H_2O, in place of the relative determinations by synchrotron X-ray scattering. The latter are based on a protein of known mol wt and reliable 280-nm absorption coefficient to determine concentrations.
5. The background is very low in D_2O buffers, even in the presence of high salt concentrations, and this permits studies of macromolecules at very low concentrations (0.5 mg/mL). D_2O is, however, a promoter of macromolecular aggregation if hydrogen bond interactions with water are important for solubility.
6. Guinier analyses at low scattering angles are not significantly affected by beam divergence or wavelength polychromicity (9–10% on D17 and D11 at the ILL). However, intensities at large Q are noticeably affected (Fig. 2), and this requires consideration in curve simulations. The situation for cameras on spallation sources is currently under assessment.

2. Preparations

2.1. Sample Preparations for Scattering

The basic requirement is for pure, monodispersed solutions of the samples at a high enough concentration for a scattering curve to be observable in the required solute–solvent contrast. For studies on a single preparation, 0.5 mL of material at 10 mg/mL is ideal for synchrotron X-ray work, and 1.5 mL at 10 mg/mL for neutron work in three contrasts. If flow or stopped-flow methods are to be used, these amounts are considerably larger. Biochemical standards of purity are adequate. Since scattered intensities are proportional to the square of the mol wt at low Q, it is essential to remove all traces of aggregates prior to measurement by gel filtration and reconcentration of the samples if the sample is prone to aggregation. Microfiltration is not sufficient. Aggregates are discovered in the course of on-line data analyses during an experimental session by nonlinear Guinier R_G plots, which curve upward at the lowest Q values. Since aggregated samples are not usable, the availability of other samples for the beamtime session as backups can be advantageous. The prior use of laser light scattering to test for the presence of aggregates has been proposed; however, in practice, this requires additional effort.

Sample dialysis prior to measurements is essential for accurate subtraction of buffer backgrounds, for which the final dialysate is employed. Slight differences in the electron density of the buffer (or exchangeable proton content for neutron work) can invalidate the subtraction. The choice of buffer will affect scattered intensities. In X-ray work, the closer a protein buffer is to pure water, the higher the sample transmission becomes, and the better the counting statistics. Phosphate buffered saline (12 m*M* phosphate, 140 m*M* NaCl, pH 7.4) is commonly used. In neutron work, a reduction of the proton content of the buffer improves the counting statistics because of the strong incoherent scattering of ^1H nuclei. Proteins are usually measured in 0, 80, and 100% D_2O buffers. Sample concentrations must be accurately known prior to measurements for mol wt and neutron matchpoint calculations. Neutron matchpoints correspond to that percentage D_2O in which the scattering density of the macromolecule is the same as that of the solvent. Protein concentrations can be measured from absorbance at 280 nm and calculating the absorption coefficient from sequences *(10)*, or by Lowry assays if absorbance cannot be measured. Assays for activity and radiation damage before and after scattering are required to demonstrate that the scattering data are meaningful; this can be important for publications.

2.2. Applications for Beamtime

Forward planning is implicit in applying for beamtime. The user liaison officer at the synchrotron or neutron facility should be contacted 6–12 mo in advance for application forms, and the names of suitable local contacts or instrument scientists. Beamtime applications are considered in competition; perhaps over twice as many days are applied for than are available. Users in the UK can contact (June 1992):

Colin Jackson (Daresbury X-ray synchrotron)
User Liaison Officer
SERC Daresbury Laboratory
Keswick Lane
Warrington
Cheshire WA4 4AD
Tel: 0925-603221
FAX: 0925-603174

Sarah Matthews (ISIS and ILL neutron beams)
Neutron Division
Building R3
Rutherford and Appleton Laboratories
Chilton
Didcot
Oxon OX11 0*Q*X
Tel: 0235-821900 Ext: 5592 or 0235-445592
FAX: 0235-445720

Herma Blank (ILL high-flux neutron reactor)
Scientific Secretariat
Institut-Laue Langevin
156X
38042 Grenoble Cedex
France
Tel: 010-33-76 48 71 11
FAX: 010-33-76 48 39 06

Users in the USA, Europe, and the Far East can obtain addresses of other national facilities from the Synchrotron Radiation Newsletter and the Neutron News, both of which are distributed free to registered users: Synchrotron Radiation Newsletter, or Neutron News, Gordon and Breach Science Publishers, Marketing Department, P.O. Box 197, London WC2E 9PX, UK.

Multiuser facilities are expensive. If purchased, the commercial cost of the beamtime is between £4000/$7000 (X-rays) and £7000/ $11,000 (neutrons)/d on the scattering camera. The large turnover of users means that users are normally held responsible for providing the sample cells and cell holders to be placed in the beam. Preregistration is advisable if use is to be made of biological laboratories at the facility. Users are held responsible for ensuring that adequate amounts of samples are ready in good time for beamtime in what is usually a tight schedule. Users are responsible for adhering strictly to local safety regulations; the facility must be informed of any biological hazards (e.g., samples purified from blood products should be obtained from HIV- and hepatitis B-free donors; samples of viruses may require safety clearance).

3. Methods

3.1. Data Collection Strategies

Data collection using any synchrotron or neutron source is best designed on the assumption that the main beam or the camera will unexpectedly stop at some point during data collection. The beam can fail because of ambient weather conditions (e.g., thunderstorms), major accident, mains electricity power surge, coolant leaks, vacuum or electronics failures, operator error, and so forth. The instrument itself can fail because of computer malfunction, jamming of beam port or sample holder, electronics breakdown, vacuum leaks, and so on. This means that all the basic data essential for subsequent data reduction should be collected before the sample and buffer runs are measured, and that the samples should be measured in order of priority. In synchrotron X-ray scattering, beam may not be available for 12–24 h periods; therefore, sessions should be booked for at least 48 h in order to be reasonably sure of collecting some data.

Scattering data are best interpreted if a satisfactory Guinier analysis can be reported. This requires a low enough experimental Q range on the camera to permit measurement in the Q range below $Q \cdot R_G$ of 1 *(1)*. For a compact globular protein, the minimum R_G value can be estimated from the expression *(3)*:

$$\log R_G = 0.365 \log M_r - 1.342 \qquad (1)$$

where M_r is the mol wt. For elongated structures, R_G values can be estimated from the overall length, L, using the approximation $R_G = L/(12)^{1/2}$ (assuming that the R_{XS} [*see below*] is low and can be neglected). Here L can be estimated from electron microscopy or from hydrodynamic simulations assuming a rigid rod model. A sufficiently large sample-detector distance is therefore needed in order to access a low enough Q value. If in doubt, it is preferable to err on the side of lower Q, even though such curves are usually more difficult to measure. For neutron scattering, a choice of wavelength λ is required. The advantage of λ at 1.0 nm or greater is that no corrections are required in absolute mol-wt calculations *(11)*.

In X-ray work, the instrument requires calibration. The *Q* range on the detector is defined using wet slightly stretched collagen, which has a diffraction spacing of 67 nm. Rat tail tendons are a ready source of collagen fibrils, requiring only a scalpel, two pairs of tweezers, and a Petri dish for extraction from a rat tail; these are stored in aqueous buffer prior to usage. Reflections are indexed based on the stronger intensities of the first, third, fifth, and ninth order peaks. The peaks should be remeasured after every beam refill. The response of the detector channels is not uniform; this is calibrated by exposure for several hours to a uniform radioactive ^{55}Fe source when there is no beam. Since the main beam diminishes in intensity during a session and the background intensity is high at low *Q*, small buffer subtraction errors occur frequently. These are minimized by measuring samples in duplicate with a buffer run in between them (all in the same cell with the same mica windows) for equal periods of time each (usually 10 or 20 min). Radiation damage is monitored by recording the data in 10 time slices during the measurement and examining the 10 subcurves for time-dependent effects. The use of 100 m*M* formate in the buffer or other additives has been proposed as a means of retarding attack by X-ray-induced free radicals *(12,13)*. Scattering curves are processed by subtracting the buffer curve from the sample curve (normalizing on the basis of ion chamber counts) and normalizing the result by dividing by the detector response curve.

In neutron work, the instrument configuration defines the *Q* range. The basic runs to be measured *(8)* are cadmium (for neutron and electronic noise background), Teflon™ (for defining the beam center and the detector area masked by the beamstop), a 1 mm thick H_2O sample (for detector response and absolute scale), an empty cell (background for the H_2O sample), and an empty cell holder (a check for stray reflections or scattering). The H_2O sample can usually be replaced by the buffer in 0% D_2O if this is a solution in dilute salts (e.g., phosphate buffered saline). Neutron transmission measurements (with an attenuated beam) are required for all samples and buffers to be measured, for use in mol wt *(13)* and matchpoint calculations, and to confirm that the dialysis into D_2O buffers has proceeded to completion. Transmissions vary from about 0.5 for 1 mm thick H_2O samples to about 0.9 for 2 mm thick D_2O samples, depending on the wavelength and temperature. Protein concentrations from absorbance at 280 nm are usually measured in the same quartz cell used for neutron data. Solutions are usually measured in 1 mm thick cells (0% D_2O) and 2 mm thick cells (70–100% D_2O). Sample counting times depend on the amount of D_2O in the buffer. Samples and buffers in H_2O require several hours of counting (in separate measurements of 1 h each), whereas those in D_2O require 10–30 min. The H_2O standard should be remeasured several times during a beam session to check for reproducibility. Buffer background subtractions are usually straightforward for the reason of instrumental stability. The raw scattering curves are first corrected by subtracting the cadmium background from each one. The buffer curve is subtracted from the sample curve, and then the final reduced curve is calculated by dividing this by the water background curve (corrected for the transmission of water) minus the empty cell curve *(8)*.

Instrument scientists, station masters, or local contacts are usually available for setting up the camera. However, since this has to be done for every user in a heavy beamtime schedule, users usually work on their own afterward unless a collaboration is involved. Where possible, local staff should be thanked for their efforts. Suggestions for improvements are usually best raised at users' meetings.

3.2. Analyses of Raw and Reduced Data

Guinier analyses of the scattering curves $I(Q)$ at low Q (Fig. 2) give the radius of gyration, R_G, and the forward scattered intensity, $I(0)$ *(1)*:

$$\ln I(Q) = \ln I(0) - R_G{}^2 Q^2/3 \qquad (2)$$

R_G values measure the degree of particle elongation. Molecular weights can be deduced from $I(0)/c$ values (c = sample concentration in mg/mL), either as relative values from the X-ray data or as absolute values from neutron data by referencing $I(0)$ to the incoherent scattering of water as a standard *(11,14)*. For elongated macromolecules, the corresponding cross-sectional radius of gyration, R_{XS}, and the cross-sectional intensity at zero angle $I(Q) \cdot Q_{Q \to 0}$ are obtained from curve analyses in a Q range larger than that above from:

$$\ln [I(Q) \cdot Q] = \{\ln [I(Q) \cdot Q]\}_{Q\to 0} - R_{XS}^2 Q^2/2 \quad (3)$$

Stuhrmann analyses of the R_G^2 and R_{XS}^2 values as a function of the reciprocal solute–solvent contrast $\Delta\rho^{-1}$ *(15)* report on the internal structure of the macromolecule:

$$R_G^2 = R_{G-C}^2 + \alpha_G \cdot \Delta\rho^{-1} - \beta_G \cdot \Delta\rho^{-2}$$

$$R_{XS}^2 = R_{XS-C}^2 + \alpha_{XS} \cdot \Delta\rho^{-1} - \beta_{XS} \cdot \Delta\rho^{-2} \quad (4)$$

where R_{G-C} and R_{XS-C} are the radii of gyration at infinite contrast, the terms in α_G and α_{XS} are a measure of the radial inhomogeneity of scattering density, and the terms in β_G and β_{XS} correspond to a second-order term that is a measure of the displacement of the center of scattering densities with the contrast.

After data have been reduced, it is necessary to validate the scattering curves before they can be interpreted. The curves at low Q should yield linear, reproducible Guinier plots (Fig. 2); a concentration series of the R_G and I(0)/c data should be made. Molecular weights and matchpoint values should be calculated for comparison with compositional data. Matchpoints are obtained by plotting $[(I(0)/ctT_s)]^{1/2}$ (t: sample thickness; T_s: sample transmission) as a function of the volume percentage D_2O in the buffer. Stuhrmann analyses should give linear R_G^2 and R_{XS}^2 plots, since the term in β is usually negligible. Further analytical details are given in *(1–3)*.

If sufficient data are obtained over a wide enough Q range (2 or 3 orders of magnitude), the Indirect Transformation Procedure of Glatter *(1)* can be applied to convert the scattering curve I(Q) (measured in reciprocal space; U of nm^{-1}) into the distance distribution function P(r) in real space (U of nm). About 10–20 B-spline mathematical functions are fitted to the scattering curve in order to quantify the experimental data in terms of a continuous analytical function, after which the B-splines are transformed to give P(r) (Fig. 5). This method gives an alternative calculation of the R_G and I(0) values based on the whole scattering curve. At the point at which P(r) becomes zero, the value of r gives the maximum dimension of the macromolecule. It is important in all data analyses to consider the statistical errors inherent in the determinations of R_G, R_{XS}, and I(0) values and in the calculation of P(r) and curve fitting (Figs. 5 and 6), in order to place the main conclusions from the scattering data on a sound basis.

3.3. Modeling of Scattering Curves

Modeling extends the interpretation of the scattering analyses. The use of small spheres to model the scattering curves (Figs. 5 and 6) is the most flexible and powerful method. The models can be compared quantitatively with structural data obtained using other techniques. Even though the information content of a scattering curve is limited, and unique structure determinations are not possible, the advantage of modeling is that it can rule out structures that are incompatible with the scattering curves, such as those based on an incorrect interpretation of electron micrographs. Other approaches to calculating scattering curves are based on spherical multishell models (Fig. 2) or the use of prolate or oblate ellipsoids.

For a two-density macromolecule, the full scattering curve (Fig. 6) is calculated from the Debye equation as applied to an assembly of spheres *(16)*:

$$[I(Q)/I(0)] = g(Q)\,[n_1\rho_1^2 + n_2\rho_2^2 + 2\rho_1^2 \sum_{j=1}^{m} A_j^{11}(\sin Qr_j/Qr_j) +$$

$$2\rho_2^2 \sum_{j=1}^{m} A_j^{22}(\sin Qr_j/Qr_j) + 2\rho_1\rho_2 \sum_{j=1}^{m} A_j^{12}(\sin Qr_j/Qr_j)]\,(n_1\rho_1 + n_2\rho_2)^{-2} \quad (5)$$

where the model is constructed from n_1 and n_2 spheres of different scattering densities ρ_1 and ρ_2; $g(Q) = 3(\sin QR - QR\cos QR)^2/Q^6R^6$ (the squared form factor of the spheres of radius R); A_j^{11}, A_j^{22}, and A_j^{12} are the number of distances r_j for that increment of j between the spheres 1 and 1, 2 and 2, and 1 and 2 in that order; m is the number of different distances r_j. A powerful mainframe computer is required, since the calculation of the terms in A_j and r_j is time-consuming. For neutron data, the simulated curve should be corrected for beam divergence and wavelength spread *(17,18)*.

The diameter of the small spheres has to be significantly less than the resolution of the scattering experiment (Fig. 6), so that g(Q) is independent of Q. The macromolecule is subdivided into cubes, which are then represented by the same number of overlapping spheres of the same volume. The total volume of the spheres is set equal to the dry volume of the protein as observed by crystallography *(10,19)* for neutron data, and to the dry volume plus the volume of the hydration shell

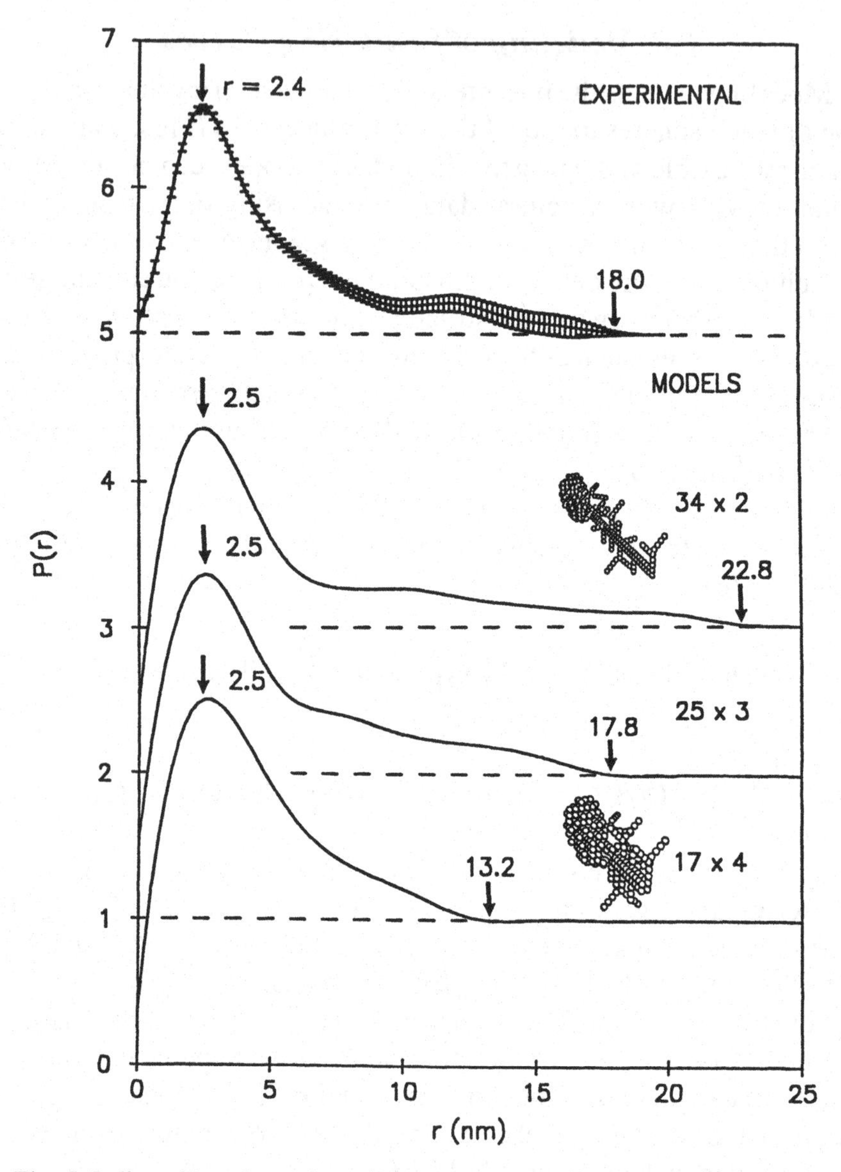

Fig. 5. Indirect Transformation of the neutron scattering curve for C1 inhibitor of complement in D_2O. The experimental P(*r*) curve corresponds to 69 I(Q) points extending to Q = 3.1 nm^{-1}, fitted using 10 B-splines and transformed using an assumed maximum length of 20.2 nm. The error margins of the transformation are shown. The experimental curve is compared with three models built from Debye spheres for the two-domain structure of C1 inhibitor (two are shown as insets, and the third is inset in Fig. 6). The *r* values corresponding to the positions of the main peak in P(*r*) and, where P(*r*) becomes zero, are arrowed. The model of length 17.8 nm (Fig. 6) gives the best agreement with experiment.

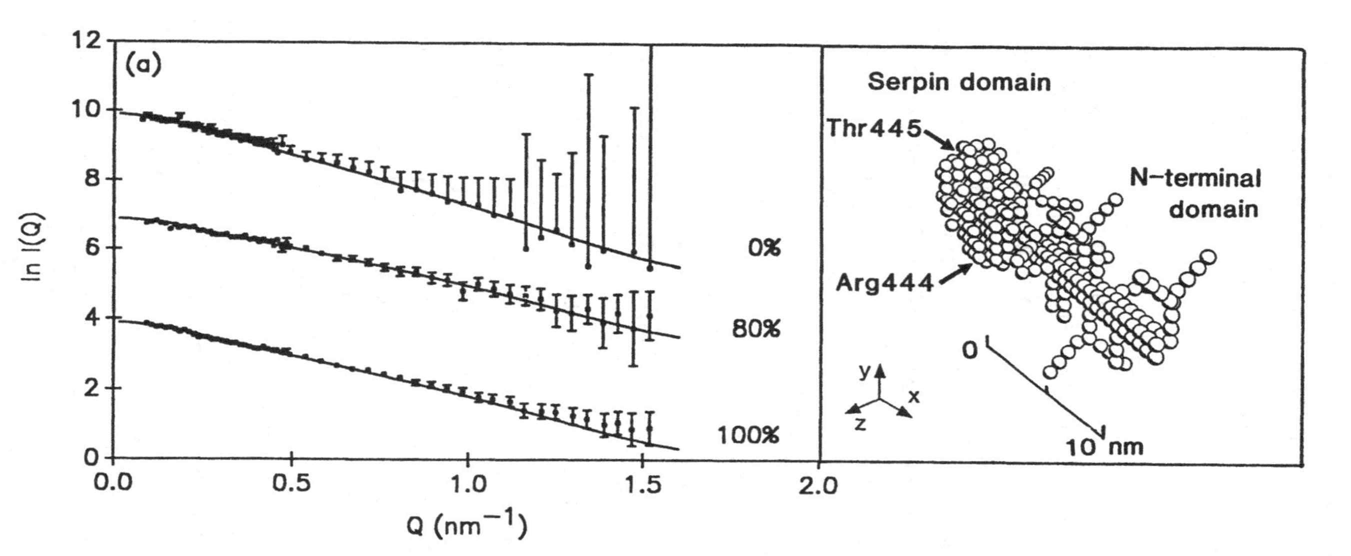

Fig. 6. Curve fitting for a Debye sphere model of C1 inhibitor. The two-domain model as shown (395 spheres of diameter 0.608 nm as shown) accounts for the scattering curves of C1 inhibitor in 0, 80, and 100% D_2O buffers, measured on D17 at the ILL Grenoble. Statistical errors in the neutron data are indicated by full (80%, 100% D_2O) or half (0% D_2O) error bars. The *N*-terminal domain is 23 spheres in length and 3 spheres in cross-section. The neutron analyses led to a improved structural model compared to those reported by electron microscopy and hydrodynamic studies.

surrounding the protein for X-ray data. A hydration of 0.3 g H_2O/g protein is usually assumed, and the volume of a protein-surface-bound (electrostricted) water molecule is 0.0245 nm^3 in place of the free water vol of 0.0299 nm^3 *(10)*. These procedures have been tested for both X-ray and neutron scattering using a known crystallographic structure *(20)*.

In the absence of a crystal structure, the sphere modeling proceeds by determining the dimensions of the simplest triaxial object that will account for the scattering curve, starting from the R_G value determined from the curve at lowest Q. If information is available on subunit or domain structures within the macromolecule from sequences or electron microscopy, these can be incorporated as constraints into the modeling. If a crystal structure is available for a subunit or domain, the scattering analyses are powerfully constrained to the extent that atomic models for the intact macromolecule can be critically assessed against the scattering data. A range of models are thus tested by a trial-and-error process against the experimental data in order to identify a family of possible structures. An example is offered by the two-domain structure of C1 inhibitor in Figs. 5 and 6. The structures of one domain, the serpin fold, is known from crystallography and solution scattering studies on α_1-antitrypsin *(20)*. The unknown structure of the other N-terminal domain was determined by solution scattering *(21)*. Two models for this structure could be rejected on the evidence of Fig. 5, to leave that shown in Fig. 6 as the final model. Another example is offered by the modeling of the 71 immunoglobulin fold domains in the structure of immunoglobulin M *(22)*. Here, crystal structures were known for all the domains. Solution scattering models were used to define allowed conformations. This study was sufficiently precise to permit the proposal of a molecular mechanism for complement activation by immunoglobulin M.

It is important to note that the resulting model only offers a curve that is equivalent to that seen experimentally. For example, the modeling of C4BP (a plasma protein) revealed that both a highly oblate ellipsoid and a more reasonable seven-armed structure fitted the experimental data equally well *(23)*. The latter was selected as the final model for reason of electron microscopy data. It is also important to confine the total number of parameters used to define the model to within that total available experimentally from the observed Q range *(24)*.

The scattering model can be compared with the modeling of frictional coefficients based on hydrodynamic spheres *(20–23,25)*. This acts as a further control of the modeling, since this is a fully independent physical measurement. This will check that the R_G from scattering is compatible with experimental sedimentation or diffusion coefficients.

References

1. Glatter, O. and Kratky, O. (eds.) (1982) *Small Angle X-ray Scattering.* Academic, New York.
2. Perkins, S. J. (1988) Structural studies of proteins by high-flux X-ray and neutron solution scattering. *Biochem. J.* **254,** 313–327.
3. Perkins, S. J. (1988) X-ray and neutron solution scattering. *New Compr. Biochem.* **18B (Part II)** 143–264.
4. Bellamy, M. F., Nealis, A. S., Aitken, J., Bruckdorfer, K. R., and Perkins, S. J. (1989) Structural changes in oxidised low density lipoproteins and the effect of the antiatherosclerotic drug probucol by synchrotron X-ray and neutron solution scattering. *Eur. J. Biochem.* **183,** 321–329.
5. Nave, C., Helliwell, J. R., Moore, P. R., Thompson, A. W., Worgan, J. S., Greenall, R. J., Miller, A., Burley, S. K., Bradshaw, J., Pigram, W. J., Fuller, W., Siddons, D. P., Deutsch, M., and Tregear, R. T. (1985) Facilities for solution scattering and fibre diffraction at the Daresbury SRS. *J. Appl. Crystallogr.* **18,** 396–403.
6. Towns-Andrews, E., Berry, A., Bordas, J., Mant, G. R., Murray, P. K., Roberts, K., Sumner, I., Worgan, J. S., Lewis, R., and Gabriel, A. (1989) A time-resolved X-ray diffraction station: X-ray optics, detectors and data acquisition. *Rev. Scient. Instrum.* **60,** 2346–2349.
7. Ibel, K. (1976) The neutron small-angle camera D11 at the high-flux reactor, Grenoble. *J. Appl. Crystallogr.* **9,** 269–309.
8. Ghosh, R. E. (1989) *A Computing Guide for Small Angle Scattering Experiments.* Institut Laue Langevin Internal Publication 89GH02T.
9. Heenan, R. K., Osborn, R., Stanley, H. B., Mildner, D. F. R., Furusaka, M., and King, S. M. (1992) The small-angle diffractometer LO*Q* at the ISIS pulsed neutron source. *J. Appl. Crystallogr.* in preparation.
10. Perkins, S. J. (1986) Protein volumes and hydration effects: the calculation of partial specific volumes, neutron scattering matchpoints and 280 nm absorption coefficients for proteins and glycoproteins from amino acid sequences. *Eur. J. Biochem.* **157,** 169–180.
11. Jacrot, B. and Zaccai, G. (1981) Determination of molecular weight by neutron scattering. *Biopolymers* **20,** 2413–2426.
12. Zipper, P., Wilfing, R., Kriechbaum, M., and Durchschlag, H. (1985) A small-angle X-ray scattering study on pre-irradiated malate synthase. The influence of formate, superoxide dismutase and catalase on the X-ray induced aggregation of the enzyme. *Z. Naturforsch.* **40c,** 364–372.

13. Durchschlag, H. and Zipper, P. (1988) Primary and post-irradiation inactivation of the sulfhydryl enzyme malate synthase: correlation of protective effects of additives. *FEBS Lett.* **237,** 208–212.
14. Kratky, O. (1963) X-ray small angle scattering with substances of biological interest in diluted solutions. *Progr. Biophys. Chem.* **13,** 105–173.
15. Ibel, K. and Stuhrmann, H. B. (1975) Comparison of neutron and X-ray scattering of dilute myoglobin solutions. *J. Mol. Biol.* **93,** 255–266.
16. Rol'bin, Yu. A., Kayushina, R. L., Feigin, L. A., and Shchredin, B. M. (1973) Calculation of the small-angle X-ray scattering intensity on a computer using a macromolecule model. *Kristallografiya* **18,** 701–705.
17. Cusack, S. (1981) Instrumental effects on the scattering curves (Appendix). *J. Mol. Biol.* **145,** 539–541.
18. Perkins, S. J. and Weiss, H. (1983) Low resolution structural studies of mitochondrial ubiquinol-cytochrome c reductase in detergent solutions by neutron scattering. *J. Mol. Biol.* **168,** 847–866.
19. Chothia, C. (1975) Structural invariants in protein folding. *Nature* **254,** 304–308.
20. Smith, K. F., Harrison, R. A., and Perkins, S. J. (1990) Structural comparisons of the native and reaction centre cleaved forms of α_1-antitrypsin by neutron and X-ray solution scattering. *Biochem. J.* **267,** 203–212.
21. Perkins, S. J., Smith, K. F., Amatayakul, S., Ashford, D., Rademacher, T. W., Dwek, R. A., Lachmann, P. J., and Harrison, R. A. (1990) The two-domain structure of the native and reaction centre cleaved forms of C1 inhibitor of human complement by neutron scattering. *J. Mol. Biol.* **214,** 751–763.
22. Perkins, S. J., Nealis, A. S., Sutton, B. J., and Feinstein, A. (1991) The solution structure of human and mouse immunoglobulin IgM by synchrotron X-ray scattering and molecular graphics modelling: a possible mechanism for complement activation. *J. Mol. Biol.* **221,** 1345–1366.
23. Perkins, S. J., Chung, L. P., and Reid, K. B. M. (1986) Unusual ultrastructure of complement component C4b-binding protein of human complement by synchrotron X-ray scattering and hydrodynamic analysis. *Biochem. J.* **233,** 799–807.
24. Luzzati, V. and Tardieu, A. (1980) Recent developments in solution X-ray scattering. *Annu. Rev. Biophys. Bioeng.* **1,** 529–552.
25. Perkins, S. J. (1989) Hydrodynamic modeling of complement, in *Dynamic Properties of Biomolecular Assemblies* (Harding, S. E. and Rowe, A. J., eds.), Royal Society of Chemistry, London, pp. 226–245.

CHAPTER 5

Determination of Macromolecular Homogeneity, Shape, and Interactions Using Sedimentation Velocity Analytical Ultracentrifugation

Stephen E. Harding

1. Introduction

Since its inception by Svedberg and coworkers in the 1920s, the analytical ultracentrifuge has provided a powerful tool in biochemistry and molecular biology, with applications ranging from simple purity checks and particle shape determinations from sedimentation velocity, isolation and purification of macromolecules using density gradient techniques right through to the evaluation of mol wt and mol-wt distributions, thermodynamic second virial coefficients, and association constants using sedimentation equilibrium, *without* the need for calibration standards. By the 1960s, the familiar "Beckman Model E" had become commonplace in biochemical laboratories worldwide.

With the advent of the "high-resolution" techniques of X-ray crystallography and nuclear magnetic resonance (NMR) for structural analysis and the relatively straightforward techniques of gel electrophoresis and gel permeation chromatography for purity and mol-wt analysis, the technique suffered a serious decline in popularity in biochemistry and molecular biology with the result that, by 1980, relatively few laboratories had this facility. However, for both structural and mol-wt analysis, the technique is now undergoing a form of renaissance, culminating in the launch of the new Optima XLA

From: *Methods in Molecular Biology, Vol. 22: Microscopy, Optical Spectroscopy, and Macroscopic Techniques* Edited by: C. Jones, B. Mulloy, and A. H. Thomas

Analytical Ultracentrifuge from Beckman Instruments (Palo Alto, CA) *(1)*. The reasons for the revival of analytical ultracentrifugation for mol-wt analysis are outlined in Chapter 6 in relation to the sedimentation equilibrium technique. For *conformational* analysis, this revival is probably the result of a realization among molecular biologists that: (1) not all biological macromolecules can be crystallized; many are available in too small a quantity (e.g., some newly engineered proteins) or are not presently amenable to full structural analysis (for example, intact, immunologically active antibodies). (2) The requirement of very high concentrations for NMR analysis of biological macromolecules—particularly those with a mol wt, M > 10,000—can lead to serious difficulties in data interpretation. Solution techniques like sedimentation analysis or X-ray scattering (*see* Chapter 4), although of low resolution, may therefore represent for many systems the only realistic "handle" on macromolecular conformation in solution. This chapter is concerned with *sedimentation velocity* analytical ultracentrifugation (i.e., where the speeds are sufficiently high to cause the macromolecule to sediment, and the nature of the form and movement of the boundary between solution and solvent is used to deduce useful information about the macromolecular system).

The basic principle of the technique is as follows: A solution of the macromolecule is placed in a specially designed sector shaped cell with transparent end windows. A mercury arc or similar light source positioned below the rotor transmits light via a monochromator or filter through the solution and then other appropriate optical components. The moving boundary can then be recorded either on photographic film, on chart paper (Fig. 1), or as digital output. For further details of the method, we still refer the reader to the classical texts of Tanford *(2)* and Schachman *(3)*. A useful introductory text is Bowen *(6)*. Introductory chapters are to be found in Price and Dwek *(7)*, van Holde *(8)*, and Rowe *(9)*.

Fig. 1. *(opposite page)* Optical records of sedimentation velocity in the analytical ultracentrifuge obtained from the MSE Centriscan. The direction of sedimentation in each case is from left to right. Top: Sedimenting boundary for a protein (methylmalonyl mutase) recorded using scanning absorption optics. Monochromator wavelength = 295 nm; scan interval = 9 min; rotor speed = 44,000 rpm; temperature = 20.0°C; Loading concentration, c° ~0.7 mg/mL; $s_{20} = (7.10 \pm 0.04)\ S$.

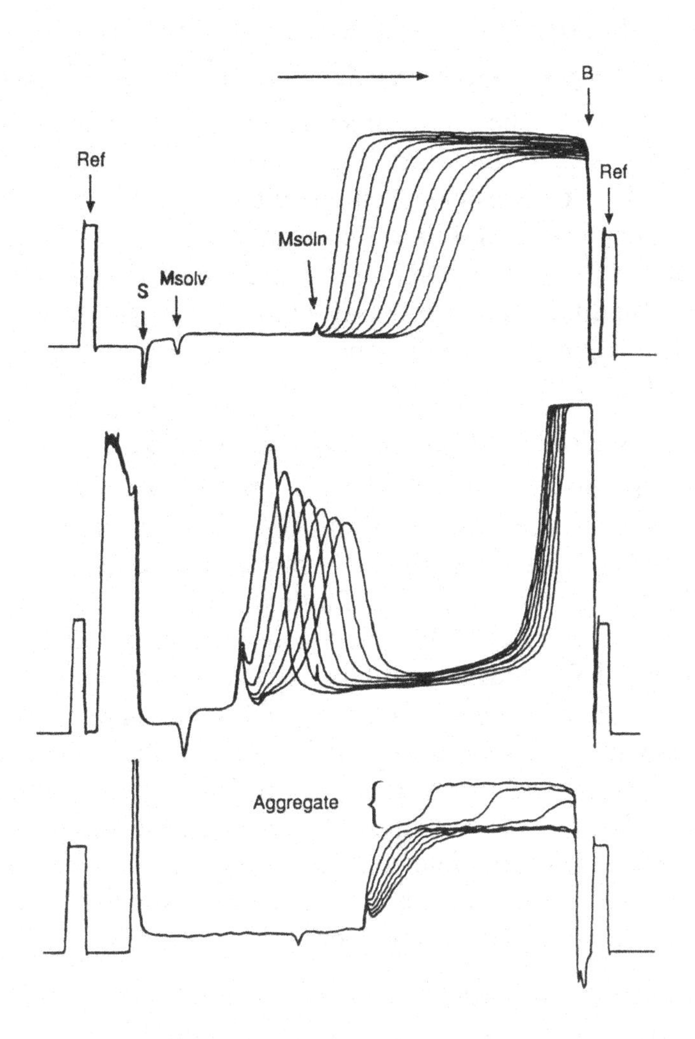

Ref: Reference marks, allowing calibration of abscissa positions in terms of actual radial displacements from the center of the rotor; S: start of cell position; B: cell base; M_{solv}: meniscus position in the solvent cell; M_{soln}: meniscus position in the solution cell. Redrawn from ref. *4*. Middle: Sedimenting boundary for a polysaccharide (heat-treated sodium alginate) recorded using scanning schlieren optics. Monochromator wavelength = 546 nm; scan interval = 30 min; rotor speed = 49,000 rpm; temperature = 20.0°C; c°~5.0 mg/mL; s_{20} = (1.22 ± 0.05) S. Lower: Sedimenting boundaries for a DNA binding protein (Gene 5) recorded using scanning absorption optics. Monochromator wavelength = 278 nm; scan interval = 8 min; rotor speed = 40,000 rpm; temperature = 20.0°C; c° ~0.7 mg/mL; $s^{\circ}_{20,w}$ = (35.5 ± 1.4) S (faster boundary) and (2.6 ± 0.1) S (slower boundary). Redrawn from ref. *5*.

2. Summary of Information Available

2.1. Homogeneity of the Macromolecular Solute

For homogeneity checks, sedimentation velocity can be used to detect the presence of impurities/polydispersity (i.e., components of different mol wt or density not in chemical equilibrium with each other) and the presence of self-association phenomena for a range of concentrations (for example, for a protein, from typically 0.2 mg/mL up to the solubility limit). It is a useful tool for assaying whether the protein solutions are still homogeneous in the very high concentrations often used for NMR.

2.2. Sedimentation Coefficient Evaluation

For the determination of sedimentation coefficients, $s^{\circ}_{20,w}$ we measure the rate of movement of the sedimenting boundary, recorded using refractometric (classical "schlieren" optics) or scanning absorption optics. The $s^{\circ}_{20,w}$ value by itself is of little interest these days, as is the traditional practice of quoting "frictional ratios" and corresponding equivalent ellipsoid of revolution "axial ratios." The necessary hydrodynamic theories have now been developed, so that the structures of complex macromolecules can be modeled using the sedimentation coefficient and related parameters *(10)*. Sedimentation coefficients have recently been used to distinguish between possible solution models for the immunological complement system *(11)* and *intact, immunologically active* antibodies for which no high-resolution structural information from X-ray crystallography or NMR is available *(12)*.

2.3. Frictional Ratio and Derived Parameters

The frictional ratio, f/f_o, can be calculated from the sedimentation coefficient and associated parameters *(see* Note 6*)*. Theoretical representations are available linking this with the axial ratios of ellipsoids (both ellipsoids of revolution and general triaxial ellipsoids) and also "bead model" representations for representing the solution conformation of relatively rigid complex structures (if combined with other solution techniques).

2.4. Flexibility Parameters

Flexibility parameters include the contour length, L, the persistence length, a, and the characteristic ratio, C_{∞}. These, particularly the ratio of *L/a*, can be useful for representing the conformations of linear biopolymers, such as nucleic acids and polysaccharides *(13)*.

2.5. Molecular Weight, M

The sedimentation coefficient can be used to give an estimate for M directly, after assumptions concerning the conformation. Alternatively, an absolute estimate can be obtained by using the sedimentation coefficient with the translational diffusion coefficient or (to a good approximation) the intrinsic viscosity, to eliminate the effects of particle conformation.

2.6. Assay for Self-Association Behavior

For self-associating or other interacting systems, simple qualitative assays (i.e., detecting whether a system is self-associating or not) and also more complicated quantitative representations in terms of interaction parameters are possible. (For details of all of Sections 2.1.–2.6., the reader is referred to ref. *13)*.

3. Availability of Instrumentation

The instrument that is still almost synonymous with analytical ultracentrifugation is the Beckman Model E, although others, such as the MSE (Crawley, UK) Centriscan, have proven popular. Many interesting adaptations of these "classical" machines have been described *(see,* for example, ref. *13)*, and many of these ideas are being incorporated into the Beckman Optima XLA Analytical Ultracentrifuge with the facility for full "on-line" analysis of the data *(1)*.

The Beckman Model E is no longer commercially available, nor is the MSE Centriscan. Second-hand instruments are usually available, but because of the complexity of the instrumentation, the molecular biologist, unless he or she has direct access to the new Beckman Optima XLA ultracentrifuge, is probably advised to consult centers where the expertise/instrumentation is available (e.g., in the US, the National Ultracentrifuge Facility at Storrs, CT, and in the UK, the Nottingham/Leicester Joint Centre for Macromolecular Hydrodynamics).

As far as the author is aware, the Model 3180 Analytical Ultracentrifuge, from the Hungarian Optical Works MOM (Budapest) is also still commercially available, at the time of going to press.

4. Materials

4.1. Choice of Solvent

It is usual to perform measurements on solutions of biological macromolecules in the presence of a low-mol-wt electrolyte at an appropriate ionic strength (typically 0.1*M* for a protein) to suppress charge effects.

4.2. Concentration/Volume Requirements of the Macromolecular Solute

If schlieren optics are to be used with ~10 mm optical path length cells, a minimum amount of 2.0 mg/mL (and ~0.4 mL) is required (lower if long [≥20 mm] optical path length cells are available); for nucleic acids and some proteins, reasonable estimates for the sedimentation coefficient may be possible for concentrations as low as 0.1 mg/mL using the UV absorption optical system. The reader may appreciate here that, compared to sodium dodecyl sulfate (SDS) gel electrophoresis or gel permeation high-pressure liquid chromatography (HPLC), the technique usually requires more material (of the order of milligrams), but much smaller amounts than, for example, NMR or X-ray crystallography. This can represent an important factor when deciding on an appropriate method for the characterization of a newly engineered protein not available in large quantities.

5. Methods

5.1. Choice of Optical System

The appropriate optical system has to be chosen, either absorption optics at an appropriate wavelength, if your macromolecule has a suitable chromophore (for proteins, conventionally ~280 nm, nucleic acids, 256 nm), or schlieren (refractive index gradient) optics, if concentrations are sufficient (usually ≥2.0 mg/mL) and if no suitable chromophore is present.

5.2. Choice of Appropriate Speed

For a globular protein of sedimentation coefficient ~2 Svedbergs (S, where $1S = 10^{-13}\,s$), a rotor speed of 50,000 rpm will give a measurable set of optical records after some hours. For larger macromolecular systems (e.g., 12S globulins or 30S ribosomes), speeds of <30,000 rpm are appropriate.

5.3. Choice of Appropriate Temperature

The standard temperature at which sedimentation coefficients are quoted is now 20.0°C (sometimes 25.0°C). If the macromolecule is thermally unstable, and depending on the nature of the macromolecule and the conditions chosen (a sedimentation velocity run can take between 1 and 12 h), temperatures down to ~4°C can be used without difficulty.

5.4. Sedimentation Coefficient Measurement

If a sedimentation coefficient (symbol "s") is desired, repeat for several concentrations, *c*, unless a low enough concentration can be used so that concentration effects are negligible (usually ≤0.5 mg/mL for a protein, but care has to be expressed if there is the possibility of association/dissociation phenomena).

5.5. Sedimentation Coefficient Calculation

The *s* value is obtained as the rate of movement of the boundary per unit gravitional field: $s = (dr/dt)/\omega^2 r$, where *r* is the radial position of the "2nd moment" of the boundary (effectively the boundary center for most applications) and ω the angular velocity in rad/s.

5.6. Correlation to Standard Conditions

For each concentration, correct the sedimentation coefficient to standard conditions (water at 20°C): symbol "$s_{20,w}$" using formulae given in, e.g., Tanford *(2)*. Knowledge of a parameter known as the "partial specific volume" (essentially the reciprocal of the anhydrous macromolecular density) is needed; this can usually be obtained from standard tables or, for proteins, can be calculated from amino acid composition data *(14)*.

5.7. Extrapolation to Zero Concentration

Plot $s_{20,w}$ vs *c* (the latter corrected for "average radial dilution" because of the sector design of the cell centerpiece) and extrapolate (usually linearly) to zero concentration (Fig. 2) to give a parameter, $s^{\circ}_{20,w}$, which can be directly related to the frictional properties of the macromolecule (the so-called "frictional ratio") and from which size and shape information may be inferred. If the macromolecule is very asymmetric or solvated, plotting $1/s_{20,w}$ vs *c* generally gives a more useful extrapolation *(9)*.

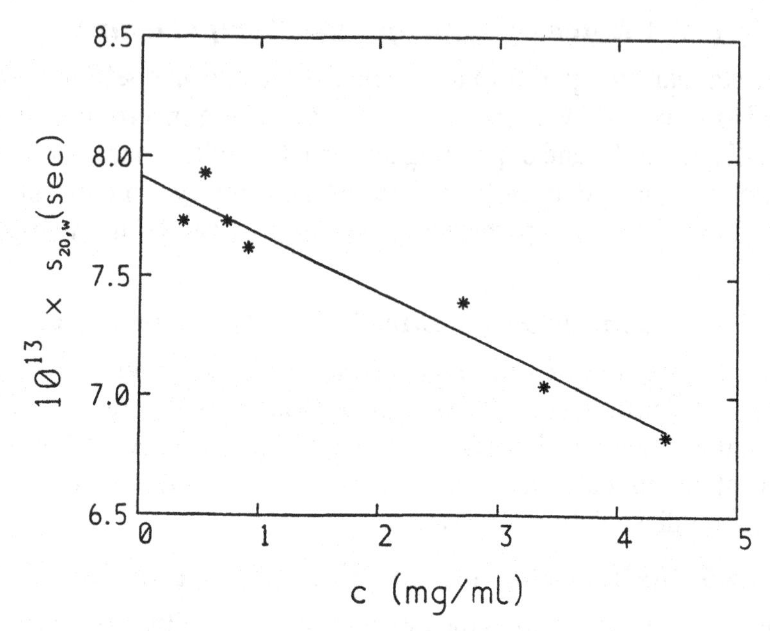

Fig. 2. Sedimentation coefficient $s_{20,w}$ vs concentration plot for an antibody (Rat IgE). $s^{o}_{20,w}$ = (7.92 ± 0.06) S.

6. Notes

1. How do we infer homogeneity? The presence of more than one sedimenting boundary (Fig. 1, bottom) can demonstrate either polydispersity or self-association phenomena. However, the converse is not necessarily true: A single sedimenting boundary is not in itself *conclusive* proof of sample homogeneity *(15)*.
2. The downward slope in a plot of $s_{20,w}$ vs concentration is a result of nonideality behavior and is characterized by the parameter k_s in the equation:

$$s_{20,w} = s^{o}_{20,w} (1 - k_s c) \tag{1}$$

3. k_s (mL/g) itself is quite a useful parameter. The ratio $k_s/[\eta]$ (where $[\eta]$ is the intrinsic viscosity [mL/g] of the macromolecule), the so-called "Wales-van Holde ratio," is a function of macromolecular conformation having a value of ~1.6 for spheres and random coils, and somewhat less for extended conformations (*see*, e.g., *16*).
4. $s^{o}_{20,w}$ can be combined with the translational diffusion coefficient (again corrected to standard conditions), $D^{o}_{20,w}$—the latter usually obtained from quasi-elastic light scattering (QLS) measurements (*see* Chapter 8)—to give an absolute value for the mol wt, M, via the Svedberg equation *(2)*.

5. For a polydisperse material, if $s^\circ_{20,w}$ is a weight average and $D^\circ_{20,w}$ from QLS a z-average, the M will be a *weight average* (M_w). It is also possible to evaluate a sedimentation coefficient distribution and, if a reasonable assumption about the conformation can be made, a mol-wt distribution *(17)*.
6. If M is known (e.g., for a protein from amino acid sequence data), together with a good idea of the "hydration" of the macromolecule (i.e., a measure of the amount of solvent—both chemically bound and physically entrapped—associated with the macromolecule), $s^\circ_{20,w}$ can provide a useful handle on macromolecular conformation.

   Specifically, $s^\circ_{20,w}$ can be used to evaluate the "frictional ratio," f/f_o (where f is the frictional coefficient and f_o is the frictional coefficient of an anhydrous* spherical particle having the same mass and $\bar{v}$ as the macromolecule under consideration):

$$[(f/f_o)] = \{M(1 - \bar{v}\rho)]/[N_A \cdot (6\pi\eta_o s^\circ_{20,w})]\}[(4\pi N_A/3\bar{v}M)]^{1/3} \quad (2)$$

   N_A is Avogadro's number, and η_o is the solvent viscosity (in this case water at 20.0°C); f/f_o, in turn, is a function of the hydration of the macromolecule, w, and the conformation via a particle shape factor known as the "Perrin function," P (or "frictional ratio owing to shape"):

$$[(f/f_o)] = P \cdot [(w/\bar{v}\rho_o) + 1]^{1/3} \quad (3)$$

   If the macromolecule is fairly rigid, P can be related directly to the axial dimensions of the macromolecule using ellipsoid of revolution or general triaxial ellipsoid representations of the data *(18)*, or perhaps more usefully, by representing the solution structure of the macromolecule in terms of arrays of spherical beads, its measurement enables complex molecules like antibodies or bacteriophages to be modeled *(10)*. An example of this is shown for two forms of a T-even bacteriophage in Fig 3. It should be stressed that, realistically, for complex modeling of this sort, a good starting estimate for the solution conformation has to be known (from, e.g., electron microscopy or X-ray crystallography) because of uniqueness problems, and for *any* type of modeling using $s^\circ_{20,w}$, a reasonable assumption about the degree of "hydration" of the protein has to be made *(10)*. A survey of 21 proteins *(19)* has shown that, for proteins, values for the hydration can vary quite markedly (for this range, $w = (0.53 \pm 0.26)$ g H_2O/g protein). For some glycoproteins,

*In the literature, the convention is sometimes chosen so that f_o refers to the *hydrated* spherical particle: In this case Eqs. (2) and (3) have to be modified accordingly, but both treatments are of course equivalent.

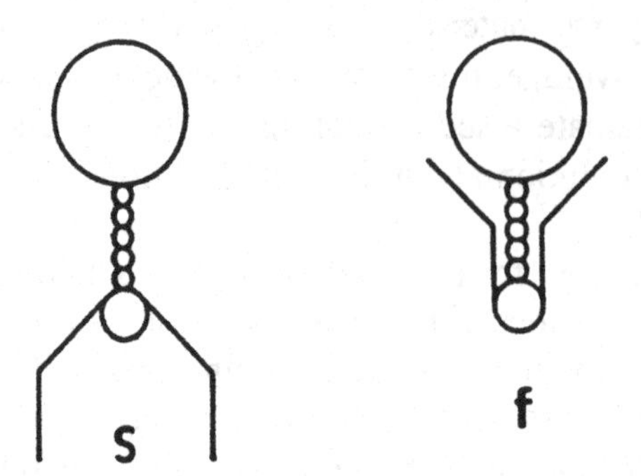

Fig. 3 Hydrodynamic bead model for a T-even bacteriophage based on sedimentation coefficient and translational diffusion coefficient data on slow (s) and fast (f) forms ($s^{\circ}_{20,w}$ ~710 S and 1020 S, respectively). Only two of the six tailfibers (each of which is modeled by 64 small beads) are shown. The calculated frictional ratios based on these models are in agreement with the measured frictional ratios determined from the sedimentation coefficient or translational diffusion coefficient, after allowance for hydration. (From ref. *10* and refs. cited therein).

this can be an order of magnitude higher. In this respect, the more complicated the models chosen, the more essential it is to have supporting information from further hydrodynamic, scattering, or thermodynamic data (e.g., intrinsic viscosity, rotational diffusion coefficients, radius of gyration, or the thermodynamic second virial coefficient *[18]*).

For polydisperse macromolecular systems, such as polysaccharides, proteoglycans, glycoproteins, and nucleic acids, the dependence of $s^{\circ}_{20,w}$ with M can give a useful guide to the general conformation (between the extremes of compact sphere, rigid rod, and random coil) and also provide flexibility information as indicated earlier *(see*, e.g., ref. *20)*.

7. A plot of $s_{20,w}$ vs concentration may show an initial upward tendency before dropping. The initial upward tendency if present is symptomatic of self-association/dissociation behavior. It is possible to model this behavior to obtain an association constant(*s*) *(21)*.
8. The technique can be used in a very simple way for investigating interactions in mixed solute systems (Fig. 4).

Glossary of Symbols

M, Mol wt (g/mol); M_w, Weight average mol wt; S, Svedberg unit ($1S = 1 \times 10^{-13}$ s); *s*, Sedimentation coefficient measured at a finite sedimenting (i.e., corrected for radial dilution) concentration. Unit: S or *s*; $s_{20,w}$, Sedimentation coefficient at a finite sedimenting (i.e., cor-

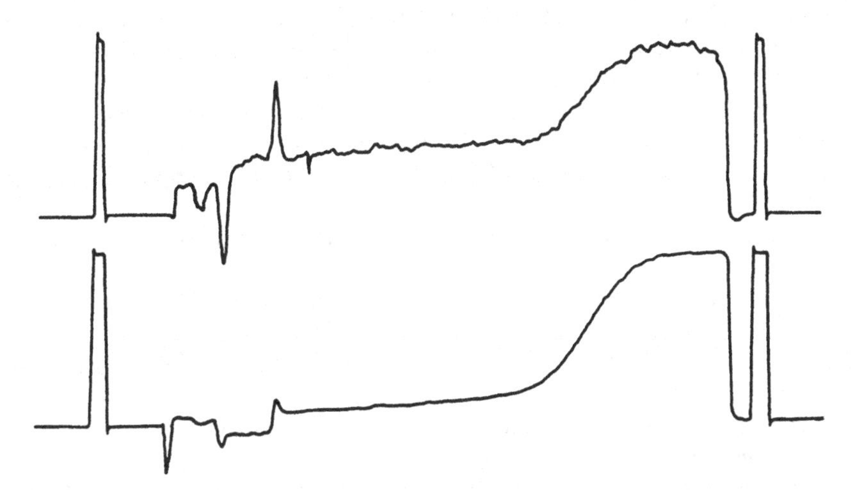

Fig. 4. Cosedimentation diagram for methyl-malonyl mutase and its cofactor (offset toward the top) scanned within 2 min of each other. The center of the sedimenting boundary is virtually the same for both, confirming that under the solvent conditions used (50 m*M*-Tris/HCl, pH 7.5 + 5 m*M* EDTA) the cofactor is bound to the protein. These scans were obtained from an MSE Centriscan, monochromator wavelength = 295 nm (bottom), 608 nm (top); rotor speed = 44,000 rpm; temperature = 20.0°C, $c° = 0.7$ mg/mL.

rected for radial dilution) concentration, *c,* and corrected to standard solvent conditions (i.e., water as solvent at a temperature of 20.0°C); $s°_{20,w}$, Infinite dilution (i.e., $c = 0$) sedimentation coefficient; $D°_{20,w}$, Infinite dilution translational diffusion coefficient (cm^2/s); k_s, Sedimentation concentration dependence regression parameter (mL/g); (f/f_o), Frictional ratio; *P*, "Perrin function" or "frictional ratio owing to shape"; *L,* Contour length (nm or cm); *a,* Persistence length (nm or cm); C_∞, Characteristic ratio; I, Ionic strength (mol/L or mol/mL); [η], Intrinsic viscosity (mL/g); $\bar{v}$, Partial specific volume (mL/g); *w*, Hydration (g H_2O/g macromolecule); ω, Angular velocity (rad/*s*); *r,* Radial displacement from the center of the rotor (cm).

References

1. Schachman, H. K. (1989) Analytical ultracentrifugation reborn. *Nature* **341,** 259–260.
2. Tanford, C. (1961) *Physical Chemistry of Macromolecules.* Wiley, New York.
3. Schachman, H. K. (1959) *Ultracentrifugation in Biochemistry.* Academic, New York.

4. Marsh, E. N., Harding, S. E., and Leadley, P. F. (1989) Subunit interactions in *Propionibacterium shermanii* methylmalonyl-CoA mutase studied by analytical ultracentrifugation. *Biochem. J.* **260,** 353–358.
5. Kneale, G. G. (1989) Hydrodynamic and fluorescence analysis of a DNA binding protein and its interaction with DNA, in *Dynamic Properties of Biomolec-ular Assemblies* (Harding, S. E. and Rowe, A. J., eds.), Royal Society of Chemistry, Cambridge, UK, pp. 171–178.
6. Bowen, T. J. (1970) *An Introduction to Ultracentrifugation.* Wiley-Interscience, London.
7. Price, N. C. and Dwek, R. A. (1978) *Principles and Problems in Physical Chemistry for Biochemists.* Clarendon, Oxford.
8. Van Holde, K. E. (1985) *Physical Biochemistry.* Prentice Hall, Englewood Cliffs, NJ.
9. Rowe, A. J. (1984) Techniques for determining mol wt. *Protein Enzyme Biochem.* **BS106,** 1–37.
10. Garcia de la Torre, J. (1989) Hydrodynamic properties of macromolecular assemblies, in *Dynamic Properties of Biomolecular Assemblies* (Harding, S. E. and Rowe, A. J., eds.), Royal Society of Chemistry, Cambridge, UK, pp. 3–31.
11. Perkins, S. J. (1989) Hydrodynamic modelling of complement, in *Dynamic Properties of Biomolecular Assemblies* (Harding, S. E. and Rowe, A. J., eds.) Royal Society of Chemistry, Cambridge, UK, pp. 226–245.
12. Gregory, L., Davis, K. G, Sheth, B., Boyd, J., Jefferis, R., Nave, C., and Burton, D. R. (1987) The solution conformations of the subclasses of IgG deduced from sedimentation and small angle x-ray scattering studies. *J. Mol. Immunol.* **24,** 821–829.
13. Harding, S. E., Rowe, A. J., and Horton, J. C. (eds.) (1992) *Analytical Ultracentrifugation in Biochemistry & Polymer Science.* Royal Society of Chemistry, Cambridge, UK.
14. Perkins, S. J. (1986) Protein volumes and hydration effects. The calculations of partial specific volumes, neutron scattering matchpoints and 280-nm absorption coefficients for proteins and glycoproteins fram amino acid sequences. *Eur. J. Biochem.* **157,** 169–180.
15. Gilbert, G. A. and Gilbert, L. (1980) Detection in the ultracentrifuge of protein heterogeneity by computer modeling, illustrated by pyruvate dehydrogenase multienzyme complex. *J. Mol. Biol.* **144,** 405–408.
16. Creeth, J. M. and Knight, C. G. (1965) On the estimation of the shape of macromolecules from sedimentation and viscosity measurements. *Biochim. Biophys. Acta* **102,** 549–558.
17. Fujita, H. (1975) *Foundations of Ultracentrifuge Analysis.* Wiley, New York.
18. Harding, S. E. (1989) Modelling the gross conformation of assemblies using hydrodynamics: The whole body approach, in *Dynamic Properties of Biomolecular Assemblies* (Harding, S. E. and Rowe, A. J., eds.), Royal Society of Chemistry, Cambridge, UK, pp. 32–56.
19. Squire, P. G. and Himmel, M. E. (1979) Hydrodynamics and protein hydration. *Arch. Biochem. Biophys.* **196,** 165–177.

20. Harding, S. E., Berth, G., Ball, A., Mitchell, J. R., and Garcia de la Torre, J. (1991) The mol wt distribution and conformation of citrus pectins studied by hydrodynamics. *Carbohydrate Polymers* **16,** 1–15.
21. Gilbert, L. M. and Gilbert, G. A. (1973) Sedimentation velocity measurement of protein association. *Methods Enzymol.* **27D,** 273–296.

CHAPTER 6

Determination of Absolute Molecular Weights Using Sedimentation Equilibrium Analytical Ultracentrifugation

Stephen E. Harding

1. Introduction

One of the most fundamental parameters describing a biological macromolecule is its mol wt, M (unit g/mol), or equivalently the dimensionless "relative molecular mass," M_r. Despite this, it is not always easy to determine or, indeed, for a heterogeneous system, define. With a homogeneous system, the simplest procedure is to calculate it directly from the chemical formula—for example, for a protein whose amino acid composition or sequence is known. In most cases, however, this simple route is not possible; the protein may be glycosylated to an extent that may be difficult to establish precisely, for example, or it may self-associate in solution. Also, the macromolecular system itself may be *polydisperse*, i.e., contain a range of macromolecular species of different mol wt, and in these systems, it is of value to determine the various types of *average* mol wt (usually the number, weight, or *z* average) or the mol-wt *distribution*. If the system is self-associating, one may also be interested in evaluating the association constant(s).

In molecular biology, the most popular methods for mol-wt determination are the so-called "relative" methods (i.e., requiring calibration using standard macromolecules of known mol wt)—namely, sodium

From: *Methods in Molecular Biology, Vol. 22: Microscopy, Optical Spectroscopy, and Macroscopic Techniques* Edited by: C. Jones, B. Mulloy, and A. H. Thomas

dodecyl gel electrophoresis (SDS-PAGE) and calibrated gel chromatography (GPC, HPLC, and so on)—*see* vol. 1, Chapters 2 and 6 in this series. Both of these methods, although relatively straightforward, have limitations for many systems. Problems may arise from nonuniform binding of SDS in the SDS-PAGE analysis of some proteins (e.g., histones, glycoproteins) or more than just the polypeptide chain mol wt may be desired.

Difficulties may occur in obtaining suitable calibration standards in gel chromatography/high-pressure liquid chromatography analysis. The increasing awareness by the biochemical and molecular biological community of these limitations is a primary reason for the resurgence of interest in absolute techniques for mol-wt measurement, such as light scattering (*see* Chapter 7) and, as we will consider here, sedimentation equilibrium in the analytical ultracentrifuge.

Although the instrumentation is essentially the same as for sedimentation velocity in the analytical ultracentrifuge (*see* Chapter 5), the basic principle of sedimentation equilibrium is somewhat different in that it is not a transport method. In a sedimentation equilibrium experiment, the rotor speed is chosen to be low enough that the forces of sedimentation and diffusion on the macromolecular solute become comparable, so that an equilibrium distribution of solute can be attained; this "equilibrium" can be established after a period of 2–96 h depending on the macromolecule, the solvent, and the run conditions. Since there is no net transport of solute at equilibrium, the recording and analysis of the final equilibrium distribution will give an *absolute* estimate for the mol-wt and associated parameters, since frictional (i.e., shape) effects are not involved. More detailed introductions to the method are given in Price and Dwek *(1)* or van Holde *(2)*, and the experienced user is referred to refs. *3–6*.

2. Summary of Information Available from This Technique

1. Absolute mol-wt values to a precision of up to ±3%.
2. For a multisubunit protein, subunit composition.
3. For heterogeneous systems, an *average* mol wt for the solute distribution: If the absorption or interference optical systems are used, the *weight average*, M^o_w, is obtained with the highest precision (where the "o" denotes over the whole solute distribution); if the schlieren optical system is used, the most directly obtainable average is the *z-average*, M^o_z.

For a given optical system, other averages can in principle be obtained, but with lower precision.

4. If the system is heterogeneous, regions nearer the base will have relatively larger amounts of higher mol-wt material than regions near the meniscus. Because of this redistribution of solute throughout the cell, the evaluation of local or "point" average mol wts, M_w, M_z, and so on (corresponding to a particular radial displacement in the centrifuge cell) is also useful, providing different information depending on the origin of the heterogeneity. (A) If the heterogeneity is owing to self-association phenomena, association constants may be determined, or if it is owing to complex formation between species of different types, interaction constants may be estimated but this is very difficult. (B) If the heterogeneity is owing to "polydispersity" (i.e., the presence of noninteracting components of different mol wt or density), mol-wt distributions may be obtained (particularly if a simple combined procedure with gel permeation chromatography can be employed *[6]*). (For further details of these, *see* ref. *7*).

3. Limitations

For large macromolecular systems, nonideality may be significant at the concentrations used, and some form of correction for thermodynamic nonideality effects may be necessary. The measured mol wts at a finite concentration will be "apparent" mol wts. Because of instability of rotor systems at very low speeds, the technique may be unsuitable for large macromolecular assemblies ($M \gtrsim 20 \times 10^6$).

4. Availability of Instrumentation

This is as for sedimentation velocity work, and is covered in Chapter 5.

5. Materials

5.1. Choice of Solvent

As for sedimentation velocity studies (*see* Chapter 5), if possible, work with a solvent of a sufficiently high ionic strength, I, to provide adequate suppression of charge effects. Such effects contribute to the thermodynamic nonideality of the system (*see* Note 4 *below*). For sedimentation equilibrium studies, it is advisable to dialyze the solution against the solvent prior to a sedimentation equilibrium run; the dialysate is used as a reference blank (although for proteins under mild conditions near their isoelectric point, only trivial errors will arise if this is not done): The reasons are related to possible redistribution phenomena of the (aqueous) solvent components (salt ions etc.) themselves (*see*, e.g., ref. *3*). Care has to be expressed with the choice of centerpiece for the centrifuge cell, since some materials are not resistant to extremes of pH, guanidine hydrodrochloride, dithiothreitol, and so forth.

5.2. Concentration and Volume Requirements for the Macromolecular Solute

This depends on the optical system being used, the path length of the cell (which must be of the "double sector" type if Rayleigh interference optics are used), and the optical properties of the macromolecule itself (refractive index or extinction characteristics): Long path length cells (20–30 mm) are generally appropriate for low concentration (0.1–1.0 mg/mL); short path length cells (10–12 mm) are for higher concentrations (≥1.0 mg/mL). To obtain a solution column length of 0.2–0.3 mm, loading volumes are typically 0.1–0.3 mL, again depending on the path length of the cell.

6. Methods

6.1. Choice of Optical System (see Chapter 5)

Rayleigh interference optics (*see* Fig. 1A for an example of the optical record) usually provide the best optical records for mol-wt analysis. For proteins and nucleic acids, the absorption optical system is, however, the most convenient (Fig. 1B), and facilitates the multiplexing of up to four samples in a run (in some cases, even more), although problems of anomalous protein absorption on cell windows have to be avoided *(8)*.

6.2. Length of Run

Smaller molecules get to sedimentation equilibrium faster than larger ones. For molecules of $M \lesssim 10{,}000$, <24 h are required; large macromolecules take 48–72 h, although for the latter, time to equilibrium can be decreased by "overspeeding," i.e., running at a higher speed for a few hours before setting to the final equilibrium speed (*see*, e.g., ref. *9*). It may, in some applications, be desirable to use shorter columns (as low as 0.5 mm); although the accuracy of the mol wts will be lower, this "short column" method offers the advantage of fast equilibrium (<24 h), which may be important if many samples need to be run and/or the macromolecule is relatively unstable.

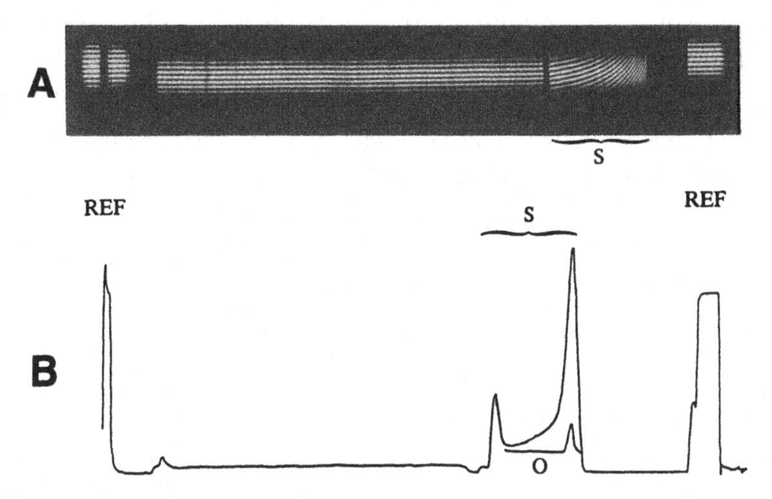

Fig. 1. Optical records of solute distributions at sedimentation equilibrium. The direction of the centrifugal field in both cases is from left to right. (A) Rayleigh interference profiles for human immunoglobulin G. (B) Absorption optical profile (280 nm) for porcine titin. REF: reference marks allowing calibration of the optical records in terms of actual distances from the rotor center. S: solution record. O: (absorption) optical baseline.

6.3. Data Capture and Analysis

If scanning absorption optics are used, the equilibrium patterns can be digitized directly on-line into a microcomputer or off-line via a graphics digitizing pattern. The average slope of a plot of Ln (absorbance) vs r^2, the square of the radial distance from the center of the rotor, will yield the weight average mol wt:

$$M = (d\mathrm{Ln}A/dr^2) \times 2RT/(1 - \bar{v}\rho)\omega^2 \quad (1)$$

where $\bar{v}$ is the partial specific volume* (typically ~0.73 mL/g for proteins, ~0.61 mL/g for saccharides), ω is the angular velocity (rad/s), and ρ the solution density (in general the solvent density ρ_0 can be used instead without giving rise to serious error in M).

With Rayleigh interference optics, the corresponding records of fringe displacement, J, vs radial displacement, r, can be obtained either

*An accurate estimate for $\bar{v}$ is normally required since, for proteins, an error of ±1% in $\bar{v}$ results in an error of ~ ±3% in M. $\bar{v}$ can be found by precision densimetry (*see* ref. *8*) or by direct calculation from the composition data (e.g., for proteins, from the amino acid sequence/composition *[10]*).

manually using "microcomparators," off-line using a laser densitometer *(11)*, or now directly on-line into a microcomputer *(12)*. An average slope of a plot of Ln[J] vs r^2 can be used in much the same way as Ln[A] vs r^2, yielding the weight average mol wt. Various other manipulations can be used to give the number and z-average mol wt *(3)*. If schlieren optics are used, the average slope of a plot of Ln[$(1/r)\cdot dn/dr$] vs r^2, where dn/dr is the refractive index gradient at a given radial position, r, yields the z-average mol wt.

7. Notes

1. Provided that an adequate baseline is available, absorption optical traces (when applicable) give a parameter (the absorbance) directly proportional to solute concentration (within the limits of the Lambert-Beer law). If the Rayleigh interference system is used, each "solution" fringe profile (cf. region "S" of Fig. 1A) corresponds to a plot of solute concentration *relative to the solute concentration at the air/solution meniscus* vs radial distance. To obtain the "absolute" fringe concentration, J, the meniscus concentration needs to be obtained. The various procedures can be found in refs. *3* and *13*. The easiest procedure—where applicable—is to run at sufficiently high speed so as to deplete the meniscus of solute. This "meniscus depletion" method *(14)* is by far the most popular for the analysis of reasonably monodisperse protein systems, although considerable caution has to be expressed, particularly when dealing with heterogeneous systems (*see*, e.g., ref. *3*)
2. For a simple monodisperse system—a situation approached by, for example, a dilute solution of a small enzyme or a polypeptide hormone, plots of Ln[A] or Ln[J] vs r^2 will be linear (Fig. 2A,B): The slope yields the mol wt as indicated earlier.
3. For a heterogeneous system, plots of Ln[A] or Ln[J] vs r^2 will be curved upward (Fig. 2C). A simple average can be taken over the whole distribution (corresponding to the "weight" average, $M_w{}^o$), or local slopes can be taken to give point average mol-wt information, such as the point weight average, M_w (Fig. 3). Obtaining the average over the whole distribution can be difficult, particularly if the optical pattern near the cell base is poorly defined: For these systems, and for the nonideal systems considered below, more advanced methods of analyzing the data are recommended (*see*, e.g., ref. *13*) rather than just trying to measure a simple slope.
4. For larger macromolecules ($M \gtrsim 100{,}000$) and/or for more concentrated solutions, nonideality may become significant, and this will tend to cause downward curvature in the Ln[A] or Ln[J] vs r^2 plots. If the

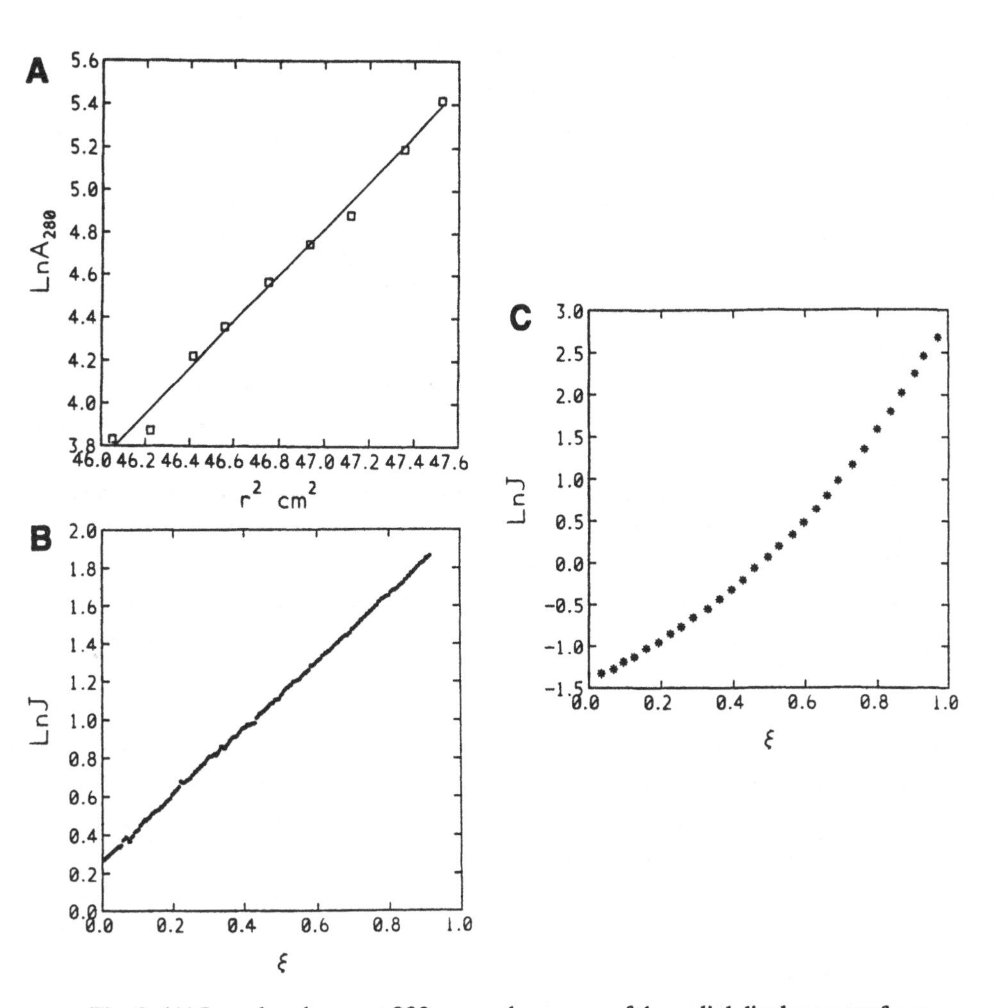

Fig. 2. (A) Log absorbance at 280 nm vs the square of the radial displacement from the center of the rotor, r^2 [for porcine titin, $M^{o}_{w} = (2.5 \pm 0.1) \times 10^6$]. (B) Log fringe concentration, J, vs a function, ξ, of r^2, normalized so it has a value of 0 at the meniscus and 1 at the cell base [$\xi = (r^2 - a^2)/(b^2 - a^2)$] for a small polypeptide, recombinant hirudin. [$M_w{}^o$ (from sedimentation equilibrium) = (7000 ± 200); M (from the amino acid sequence) = 6964.] (C) Log fringe concentration vs ξ for a glycoprotein (BM GRE) from the bronchial mucus of a chronic bronchitis patient. $M_w{}^o = (6.2 \pm 0.4) \times 10^6$ (Reproduced from ref. *15*, with permission.)

material is not significantly heterogeneous, then a simple extrapolation from a single experiment of point (apparent) mol wt to zero concentration can be made, to give the infinite dilution "ideal" value (in general, reciprocals are usually plotted—Fig. 4). Whole distribution "number" and "z"-averages, and point number and z-averages can also *in principle* be

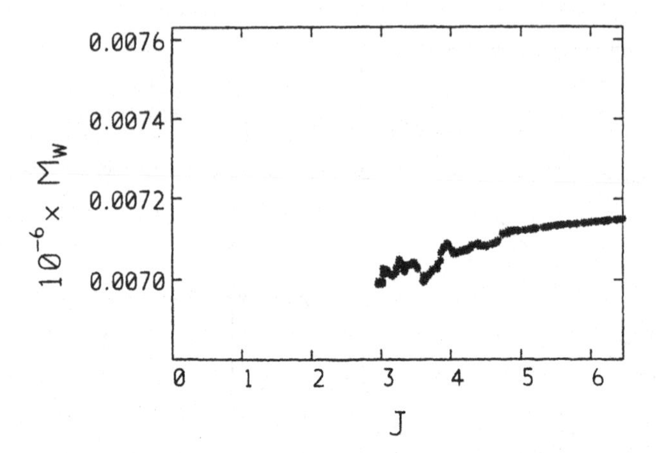

Fig. 3. Point weight average mol wt vs (fringe) concentration plot for recombinant hirudin. Note there is no evidence for dimerization behavior, i.e., M_w is ~ constant across the solute distribution.

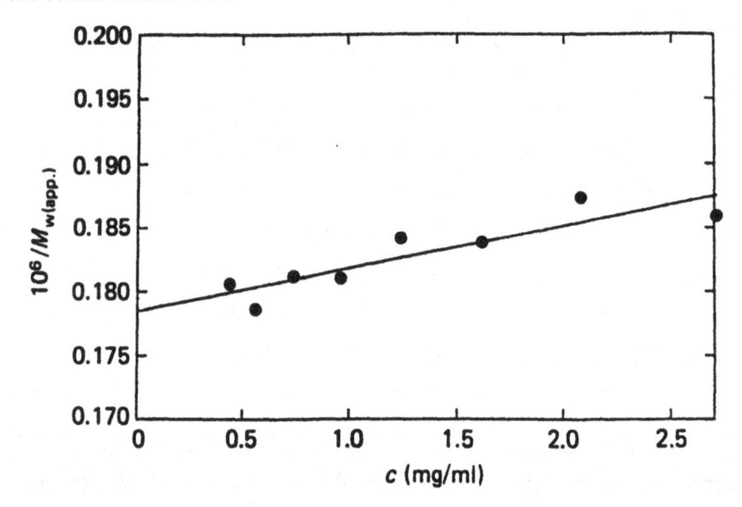

Fig. 4. Plot of the reciprocal of the (apparent) mol wt vs concentration for turnip yellow mosaic virus (TYMV). M_w [from extrapolation of point average values to zero concentration] = $(5.8 \pm 0.2) \times 10^6$; M^o_w (the weight average over the whole distribution, uncorrected for nonideality) = $(5.5 \pm 0.2) \times 10^6$. (Reproduced from ref. *16*, with permission).

obtained, although this usually requires data of the highest precision (from, e.g., on- or off-line multiple data acquisition). For this, the reader is referred to more advanced texts (e.g., ref. *7*).

5. For those heterogeneous systems where nonideality is severe, several sedimentation equilibrium experiments performed at different loading concentrations and extrapolation of "whole-cell" mol wt to zero concen-

tration are necessary. For these systems, it is worth adding that the effects of nonideality and heterogeneity can partly cancel each other out and, in some cases, yield a "pseudo-ideal" linear plot of Ln[*A*] or Ln[*J*] vs r^2, which can be misleading.

Glossary of Symbols

M, Mol wt (g/mol); M_r, Relative molecular mass; M^o_w, Weight average mol wt for the distribution of solute within a centrifuge cell (g/mol); M^o_n, Number average mol wt for the distribution of solute within a centrifuge cell (g/mol); M^o_z, *z*-Average mol wt for the distribution of solute within a centrifuge cell (g/mol); M_w, "Point" (i.e., at a given radial position in the centrifuge cell) weight average mol wt (g/mol); M_n, Point number average mol wt (g/mol); M_z, Point *z*-average mol wt (g/mol); *r*, Radial displacement from the center of the rotor (cm); *A*, Absorbance; *J*, Displacement of the Rayleigh interference fringes (corrected for any finite solute concentration at the air/solution meniscus) normal to the direction of the centrifugal field. At any given radial position in the solute distribution, *J* is directly proportional to the weight concentration, *c* (g/mL), of the solute at that radial position; ω, Angular velocity (rad/s); I, Ionic strength (mol/L or mol/mL); $\bar{v}$, Partial specific volume (mL/g); *T*, Temperature (K); *R*, Universal gas constant (8.314×10^7 erg/mol/K).

References

1. Price, N. C. and Dwek, R. A. (1978) *Principles and Problems in Physical Chemistry for Biochemists.* Clarendon, Oxford.
2. Van Holde, K. E. (1985) *Physical Biochemistry.* Prentice-Hall, Englewood Cliffs, NJ.
3. Creeth, J. M. and Pain, R. H. (1967) The determination of molecular weights of biological macromolecules by ultracentrifuge methods. *Progr. Biophys. Mol. Biol.* **17,** 217–287.
4. Teller, D. C. (1973) Characterization of proteins by sedimentation equilibrium in the analytical ultracentrifuge. *Methods Enzymol.* **27,** 346–441.
5. Fujita, H. (1975) *Foundations of Ultracentrifuge Analysis.* Wiley, New York.
6. Harding, S. E., Ball, A., and Mitchell, J. R. (1988) Combined low speed sedimentation equilibrium/gel permeation chromatography approach to molecular weight distribution analysis. *Int. J. Biol. Macromol.* **10,** 259–264.
7. Harding, S. E., Rowe, A. J., and Horton, J. C. (eds.) (1992) *Analytical Ultracentrifugation in Biochemistry and Polymer Science.* Royal Society of Chemistry, Cambridge, UK.
8. Rowe, A. J. (1984) Techniques for determining molecular weight. *Protein Enzyme Biochem.* **BS106,** 1–37.

9. Chervenka, C. H. (1969) *A Manual of Methods for the Analytical Ultracentrifuge*, Beckman Instruments, Palo Alto, CA, p. 43.
10. Perkins, S. J. (1986) Protein volumes and hydration effects. The calculations of partial specific volumes, neutron scattering matchpoints and 280-nm absorption coefficients for proteins and glycoproteins from amino acid sequences. *Eur. J. Biochem.* **157,** 169–180.
11. Rowe, A. J. and Harding, S. E. (1992) Off-line schlieren and Rayleigh data capture for sedimentation velocity and equilibrium analysis, in *Analytical Ultracentrifugation in Biochemistry and Polymer Science* (Harding, S. E., Rowe, A. J., and Horton, J. C., eds.), Royal Society of Chemistry, Cambridge, UK, pp. 49–62.
12. Laue, T. (1992) On-line data acquisition and analysis from the Rayleigh interferometer, in *Analytical Ultracentrifugation in Biochemistry and Polymer Science*, (Harding, S. E., Rowe, A. J., and Horton, J. C., eds.) Royal Society of Chemistry, Cambridge, UK, pp. 63–89.
13. Creeth, J. M. and Harding, S. E. (1982) Some observations on a new type of point average molecular weight. *J. Biochem. Biophys. Meth.* **7,** 25–34.
14. Yphantis, D. A. (1964) Equilibrium ultracentrifugation of dilute solutions. *Biochemistry* **3,** 297–317.
15. Harding, S. E. (1984) An analysis of the heterogeneity of mucins. No evidence for self-association phenomena. *Biochem. J.* **220,** 117–123.
16. Harding, S. E. and Johnson, P. (1985) Physicochemical studies on turnip yellow mosaic virus: Homogeneity, molecular weights, hydrodynamic radii and concentration dependence parameters. *Biochem. J.* **231,** 549–555.

CHAPTER 7

Classical Light Scattering for the Determination of Absolute Molecular Weights and Gross Conformation of Biological Macromolecules

Stephen E. Harding

1. Introduction

Classical light scattering, like sedimentation equilibrium in the analytical ultracentrifuge, can provide a powerful absolute method for the determination of molecular weights of macromolecules. By "classical" light scattering (as opposed to "dynamic" light scattering [cf. Chapter 8]) we mean the measurement of the total or time-integrated intensity scattered by a macromolecular solution compared with the incident intensity for a range of concentrations and/or angles; this information can be used to deduce the mol wt, M, gross conformation (from measurement of a parameter commonly referred to as the "radius of gyration," R_G), and thermodynamic nonideality parameters (particularly the thermodynamic second virial coefficient, B, which can also yield potentially useful information on molecular conformation). Classical light scattering is often referred to in the literature as "total intensity light scattering" ("TILS"), "static light scattering," "integrated light scattering," "differential light scattering," "traditional light scattering," or simply "light scattering."

Although a more rapid and, in principle, more convenient alternative to sedimentation equilibrium, the method has until relatively recently suffered greatly from the "dust problem"; viz. all solutions/scattering

From: *Methods in Molecular Biology, Vol. 22: Microscopy, Optical Spectroscopy, and Macroscopic Techniques* Edited by: C. Jones, B. Mulloy, and A. H. Thomas

cells have to be scrupulously clear of dust and supramolecular particles, particularly for the analysis of solutes of low mol wt ($\lesssim$50,000). This has resulted in many cases in the requirement for unacceptably large amounts of purified material: experiments on incompletely purified solutions have been of little value.

Two developments have made the technique now worth serious consideration by biochemists or molecular biologists: (1) the use of laser light sources, providing high collimation, intensity, and monochromaticity. These, plus the additional property of high degree of coherence also form the basis of the "dynamic" light scattering technique (*see* Chapter 8). (2) The coupling of gel permeation high-pressure liquid chromatography ("GPC") systems on-line to a light scattering photometer via the incorporation of a flow cell facilitates considerably the analysis of polydisperse materials and, more significantly, provides a very effective on-line "clarification" system from dust and other supramolecular contaminants.

Other introductory chapters on classical light scattering are to be found in van Holde *(1)* and Tanford *(2)*. More detailed treatments are given in, for example, the highly used texts of Stacey *(3)*, van de Hulst *(4)*, and Kerker *(5)*. A useful introduction to the coupling of GPC to multiangle classical (laser) light scattering ("MALLS") has been given by Jackson et al. *(6)*.

2. Summary of Information Available

1. Molecular weight (if the solution is polydisperse this will be a weight average).
2. Gross conformation information (from measurement of the radius of gyration, R_G). Measurement of the thermodynamic second virial coefficient ("B" or "A_2") can also yield gross conformational information, provided that satisfactory account can be taken of contributions to B from polyelectrolyte behavior and also from possible self-association or aggregation behavior.
3. If a coupled GPC/multiangle light scattering system is employed for a polydisperse system (e.g., a mucus glycoprotein, proteoglycan, or a polysaccharide), mol-wt distribution, R_G distribution, and from double logarithmic plots of M vs R_G and other combinations of light scattering data, additional gross conformational information can be obtained.

Classical light scattering is generally most suitable for macromolecules within the mol-wt range 50,000–50 $\times$ 10^6 g/mol.

3. Summary of Limitations

1. Sample clarification from dust and supramolecular aggregates. This is particularly serious if solutions of macromolecules of M $\lesssim$ 50,000 are being studied.
2. A separate, precise measurement of the refractive index increment, *dn/dc* is required, preferably at the same wavelength used in the light scattering photometer.
3. For very large (say $\approx$ 1 µm, M $\approx 10^{11} - 10^{12}$ g/mol) or optically dense macromolecular assemblies, the representation of the data in terms of molecular parameters can become exceedingly difficult since, because of the greater complexity of the theory involved, the so-called Rayleigh-Gans-Debye (RGD) approximation becomes inapplicable *(1)*.

4. Outline of Theory

For solutions of macromolecules or macromolecular assemblies, the basic equation for the angular dependence of light scattering is the Debye-Zimm relation:

$$Kc/R_\theta \cong \{1 + (16\pi^2 R_G^2/3\lambda^2)\sin^2[(\theta/2)]\}\,[(1/M) + 2Bc] \quad (1)$$

where it is assumed that the second virial coefficient, B, is sufficient to represent nonideality (i.e., third and higher order terms are assumed negligible). R_θ is the Rayleigh excess ratio, the ratio of the intensity, i_θ of excess light scattered (compared to pure solvent) at an angle θ to that of the incident light intensity, I_o (a cos θ correction term is needed if unpolarized light is used); K is an experimental constant dependent on the square of the solvent refractive index, the square of the refractive index increment (*dn/dc*, mL/g), and the inverse fourth power of the incident wavelength; R_G is extensively referred to as the "radius of gyration" of the macromolecule, c is the solute concentration (g/mL), and B is the thermodynamic second virial coefficient (mL · mol/g^2). If the solute is heterogeneous, M (g/mol) will be a weight average, M_w, and R_G a *z*-average. Equation (1) is generally a good representation for particles whose maximum dimension is $<\lambda$. For larger particles, much more complex representations are necessary.

For particles $\lesssim\lambda/20$ ($\equiv$M $\lesssim$ 50,000) the angular term in Eq. (1) is small: No angular dependence measurements are necessary to obtain M (although R_G cannot be obtained). When this is not the case, a double extrapolation to zero angle and zero concentration is necessary—this is usually performed on a grid-like plot referred to as a "Zimm plot" (Fig. 1) or measurement at a single angle assumed low enough so that $\sin^2(\theta/2) \approx 0$ may be adequate. Alternative methods of representing the data have in the past been in terms of "disymmetry," $z(\theta)$ (the ratio of the scattering intensity at an angle θ [typically 45°] to that at 180–θ), vs angle plots. Both $z(\theta)$ and R_G provide useful guides to the gross conformation of a macromolecule (between the extremes of sphere, rod, or random coil). Two further useful representations of the scattering data (*see*, e.g., ref. *7*) are the "Cassassa-Holtzer plot," used for example for the estimation of mass per unit lengths of rod-shape macromolecules *(8)*, and the "Kratky plot."

For fairly rigid macromolecules, R_G can be used directly to model gross conformation: (1) as an additional parameter to the sedimentation coefficient (*see* Chapter 16) and other hydrodynamic parameters for representing the structure of complex macromolecules in terms of "bead modeling" *(9)*, and (2) as a parameter, after combination with the second virial coefficient, B, and intrinsic viscosity parameters for representing structures in terms of general triaxial ellipsoids *(10)*.

5. Principal Types of Measurement

5.1. Turbidimetry

This involves the measurement of the total loss of intensity by a solution through scattering, summed over the entire angular intensity envelope, compared with the intensity of the incident radiation. This type of measurement can be performed on a good-quality spectrophotometer (whose detector does not accept appreciable amounts of scattered light). Measurements are made at wavelengths away from absorption maxima. Turbidimetry is generally suitable for macromo-

Fig. 1 (*opposite page*). Zimm Plots. The second virial coefficient, B, can be evaluated from the slope of the $\theta = 0$ line; R_G from the limiting slope of the $c = 0$ line. The common intercept is 1/M. R_θ is the scattered intensity function (the "Rayleigh excess ratio"), and K a constant depending on the square of the refractive index increment, the refractive index of the solvent, and the wavelength of the incident radiation. k is an arbitrary constant (positive or negative) chosen to "space out" the data. Note either the $c = 0$ line (usual) or the $\theta = 0$ line can have the maximum slope, depending on the relative magnitudes of R_G and B. (A) Zimm plot for a (diptheria) antigen-antibody aggregate. $k = 200$ mL/g. M ~78×10^6. (Data replotted from ref. *20*). (B) Zimm plot for a polysaccharide (*L. hyperborea* sodium alginate). $k = 500$ mL/g. M = (217,000 ± 10,000). B ~7.0×10^{-3} mL · mol / g^2. R_G ~59 nm. (Data replotted from ref. *12*.)

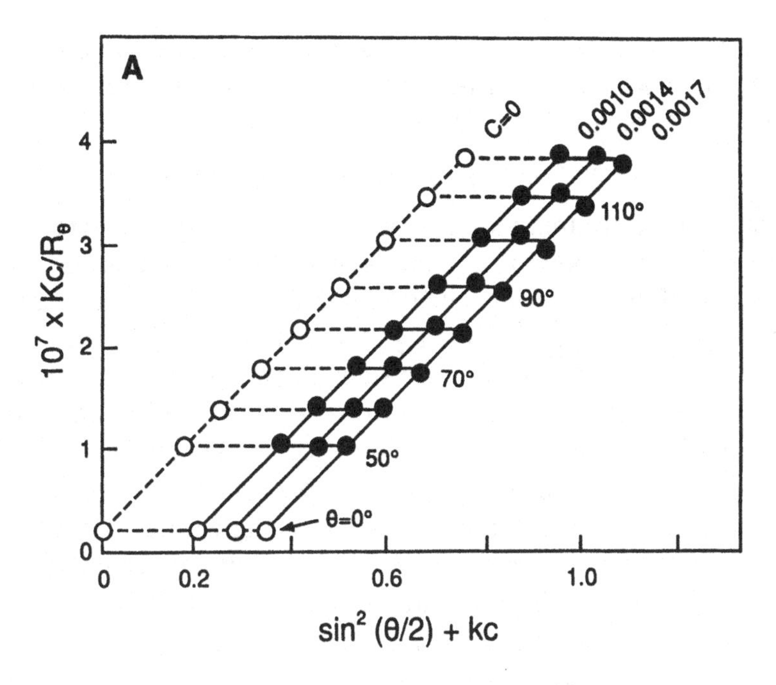
A
C=0
0.0010
0.0014
0.0017
110°
90°
70°
50°
θ=0°
$10^7 \times Kc/R_\theta$
0
1
2
3
4
0
0.2
0.6
1.0
$\sin^2(\theta/2) + kc$

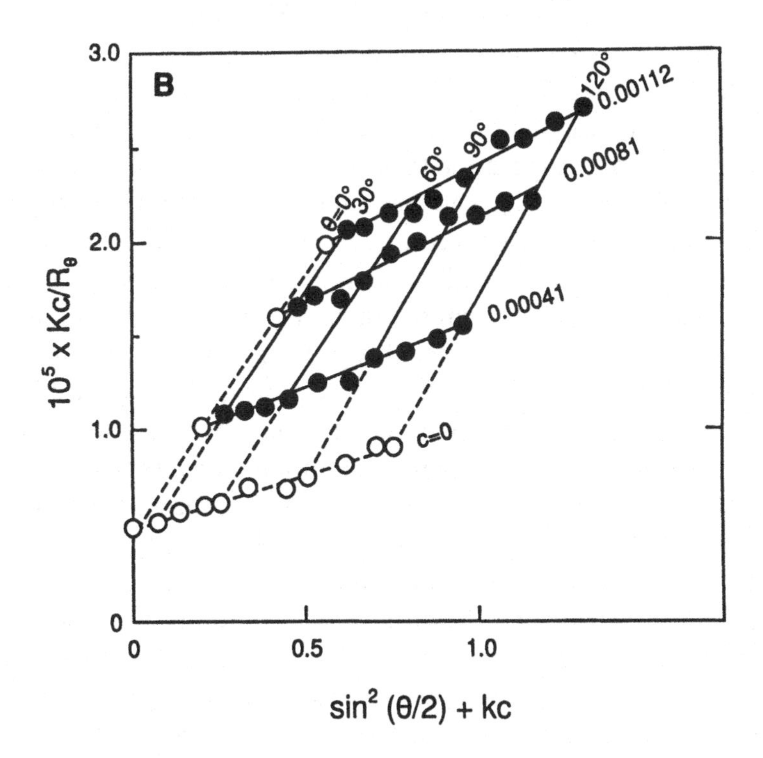
B
120°
0.00112
0.00081
90°
60°
30°
θ=0°
0.00041
c=0
$10^5 \times Kc/R_\theta$
0
1.0
2.0
3.0
0
0.5
1.0
$\sin^2(\theta/2) + kc$

lecular assemblies of $M \geq 10^5$, and has been used for the measurement of molecular weights of viruses and estimating the number concentrations of bacteria and bacterial spores (*see*, e.g., ref. *11*).

5.2. Low-Angle Light Scattering

Scattering measurements are performed at only one fixed small angle ($\leq 8°$). The angle is assumed low enough such that no angular correction of the scattering data is required, although extrapolation to zero concentration of (Kc/R_θ) may still be necessary. The method can provide values for M and B of a system, but not R_G, since no record is made of the angular dependence of Kc/R_θ. Although at low angles scattering intensities are higher and hence solute concentrations can be correspondingly lower, the dust problem is correspondingly far more severe. The on-line coupling of a low-angle (laser) light scattering, "LALLS," photometer to GPC has largely circumvented this problem and facilitates the measurement of mol-wt distributions for heterogeneous materials *(12,13)*.

5.3. Multiangle Light Scattering

This can involve a goniometer arrangement or fixed detectors at multiple angles (*see* Fig. 2A). Performing measurements at multiple angles permits extrapolation of the ratio Kc/R_θ to zero $\sin^2(\theta/2)$, which, together with an extrapolation to zero concentration, forms the basis of the Zimm plot (Fig. 1). The method can yield M, B, and R_G. Plots of (Kc/R_θ) are only linear over a wide range of angles for randomly coiled macromolecules. For globular macromolecules, "Guinier plots" of $\mathrm{Ln}(Kc/R_\theta)$ vs $\sin^2(\theta/2)$ can facilitate the angular extrapolation, or for highly branched macromolecules (for example, for some polysaccharides), "Berry" plots" $[(Kc/R_\theta)^{1/2}$ vs $\sin^2(\theta/2)]$ can be used.

A good example of the experimental arrangement involving a laser light source, how the instrument is calibrated, and application to an associative/dissociative system (hemoglobin) has been given by Johnson and McKenzie *(14)*. As with LALLS, multiangle (laser) light scattering (MALLS) photometers have been coupled on-line to GPC, facilitating the analysis of heterogeneous solutions and also largely circumventing the dust problem *(6,15)*. The simultaneous measurement at multiple angles permits the detection of possible remaining problems at low angle (through, for example, the shedding of debris from the columns). Distributions of R_G—provided the maximum dimen-

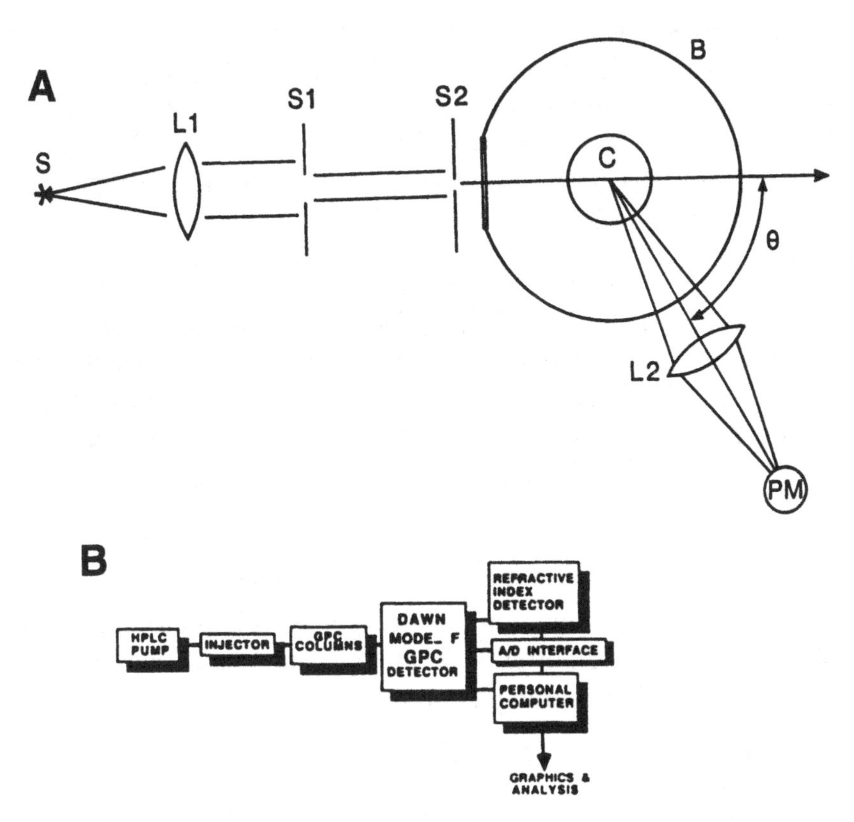

Fig. 2. Schematic light scattering systems. (A) Light scattering photometer. S light source; S1, S2 collimating apertures; L1, L2 lenses, B thermostatted bath (designed to minimize stray reflections); C cuvet (square or cylindrical); PM photomultiplier. If S is a well collimated laser, the collimating lens, L1, is not required. (Data replotted from ref. *20*.) (B) Configuration of an on-line GPC/multiangle laser light scattering photometer system. (From ref. *15*.)

sion of the macromolecule is > λ/20—as well as mol wt can be obtained, and, using the instrument in so-called batch (i.e., not coupled to GPC) mode, B can be estimated. Nonideality effects are not usually as severe for GPC-MALLS, since concentrations of volume "slices" passing through the flow cell are much smaller than the initial loading concentration: In many cases an extrapolation to zero concentration or knowledge of the second virial coefficient is not necessary to obtain a satisfactory estimate for M. One final advantageous feature for heterogeneous systems is that, since for a given M the corresponding R_G is estimated, for a homologous or quasi-homologous polymer distribution (e.g., DNA or a polysaccharide), the form of a plot of Log [M] vs Log [R_G] can provide an estimate for the gross conformation between the extremes of sphere, rigid rod, and random coil *(16)* as we have already indicated above. Examples of the use of GPC-MALLS for the determination of the mol-wt distribution of a well characterized polysaccharide are given in Fig. 3.

6. Availability of Instrumentation

This is less of a problem compared with analytical ultracentrifugation (*see* Chapters 5 and 6). A survey of recent instrumentation can be found in ref. *17*. Most of the instruments available permit dynamic as well as classical light scattering, or incorporate a flow cell and can be coupled on-line to a GPC system.

7. Materials

7.1. Choice of Solvent Media

As with sedimentation equilibrium, solutions should be dialyzed against an appropriate buffer of defined pH and ionic strength, I. For polyelectrolytes, ionic strengths of at least 0.3 are recommended.

7.2. Concentrations/Volume Requirements

This depends on:

1. The mol wt of the macromolecule, since the scattering is approximately proportional to concentration × mol wt;
2. The output of the laser; and
3. The clarity of the solutions.

For scrupulously clean solutions, a 5-mW laser for a loading concentration of 3 mg/mL is sufficient for particle mol wt as low as ≈40,000. For smaller macromolecules, a proportionately higher concentration and/or higher laser power is required. If a flow cell arrangement is used on-line to GPC, depending on the extent of the fractionation and the extent of clarification the columns can provide, higher loading concentrations may also be required. If a flow cell arrangement is used, loading volumes can be as low as 100 µL; for standard fluorimeter cuvets, up to ≈ 3 mL; if cylindrical cuvets are used, small diameters (i.e., ≤ 2 cm) are to be avoided because of extraneous scattering/reflections from the glass walls, although large-diameter cuvets can be expensive in terms of quantity of solution required.

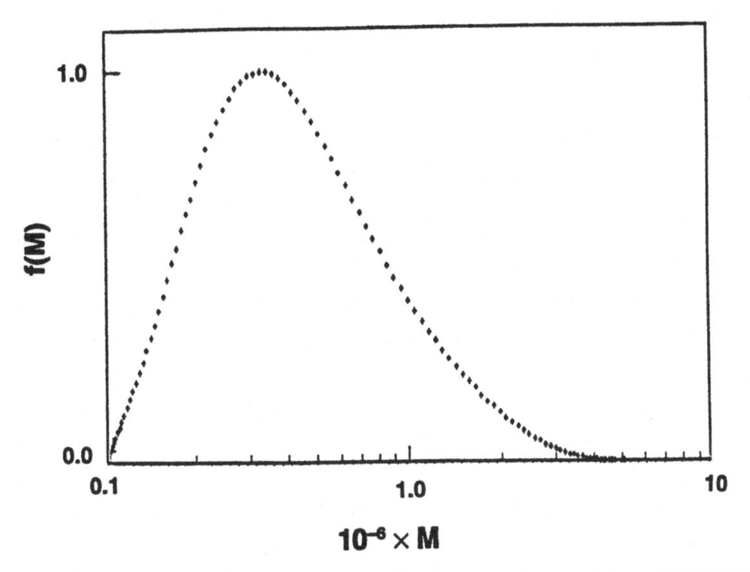

Fig. 3. Molecular-weight distribution of T-500 dextran obtained from GPC/MALLS. Weight-average mol wt for the whole distribution, M_w^o ~480,000.

8. Methods

We describe briefly here the operation of a commercially available on-line GPC-MALLS system (Fig. 2B), since this in the author's opinion appears the most versatile (for more details, *see* ref. *15*).

1. Choose GPC columns/equipment as appropriate (*see* vol. 1, Chapter 2 in this series). A pulse-free high-pressure liquid chromatography pump is essential. A guard filter upstream is desirable, as is prefiltering solutions through an appropriate millipore filter (e.g., 0.22 µm). For the Dawn F system (Wyatt Technology, Santa Barbara, CA), a ≈ 100-µL microinjection loop is desirable. A column by-pass option can be installed if fractionation is not required (viz. if Zimm plot measurements are desired for a range of loading concentrations).
2. The light scattering photometer has to be "*calibrated*" usually with a strong Rayleigh (i.e., maximum dimension $<\lambda/20$) scatterer (e.g., toluene) whose scattering properties are known (*see*, e.g., refs. *14,18*). Calibration is necessary, because the ratio of the intensities of the scattered and incident beams is usually very small ($\approx 10^{-6}$).
3. For simultaneous multiangle detection, the detectors have to be "*normalized*" to allow for (i) the different scattering volumes as a function of angle and (ii) the differing responses of the detectors. This is normally

performed using a solution of macromolecules whose M is $\lesssim 50{,}000$, or a solution of a larger macromolecule whose R_G is known (e.g., T-500 dextran).

4. A suitable concentration detector (UV absorbance for proteins/nucleic acids or a refractive index detector for polysaccharides) is incorporated downstream from the light scattering photometer. The volume delay between the light scattering photometer and the concentration detector needs to be accurately known.
5. The refractive index increment at the scattering wavelength used (and also, if appropriate, at the wavelength of the refractive index concentration detector), if not known, needs to be measured (*see* ref. *19*). Further, if the second virial coefficient, B, is not known, and if column loading concentrations are high, the Kc/R_θ ratio needs to be evaluated as a function of concentration as well as angle, and a double extrapolation to zero angle/zero concentration can be performed using a Zimm plot (*see* Fig. 1). As already mentioned, for the combined GPC/MALLS method, nonideality corrections are not usually significant, since after fractionation, the scattering concentrations are very small ($\approx$0.1 mg/mL or less).

Glossary of Symbols/Terms

TILS, Total intensity light scattering; GPC, Gel permeation chromatography; LALLS, Low-angle laser light scattering; MALLS, Multiangle laser light scattering; RGD, Rayleigh-Gans-Debye; M, Mol wt (g/mol); M_w, Weight average mol wt (g/mol); B or A_2, Second thermodynamic (or "osmotic pressure") virial coefficient (mL · mol/g^2); R_G, Root mean square radius about the center of mass ("Radius of gyration") (nm or cm); θ, Scattering angle; $z(\theta)$, Disymmetry ratio; *n*, Refractive index; *c*, Solute concentration (g/mL); *dn/dc*, Refractive index increment (mL/g); λ, Wavelength of the incident light (nm or cm); R_θ, Rayleigh excess ratio; *K*, Experimental constant (mL · mol/g^2); I_o, Intensity of incident light; i_θ, Excess scattered light intensity from a solution (compared to pure solvent) at an angle θ.

References

1. Van Holde, K. E. (1985) *Physical Biochemistry*. Prentice-Hall, Englewood Cliffs, NJ.
2. Tanford, C. (1961) *Physical Chemistry of Macromolecules*. Wiley, New York.
3. Stacey, K. A. (1956) *Light Scattering in Physical Chemistry*. Academic, New York.

4. Van de Hulst, H. C. (1957) *Light Scattering by Small Particles.* Wiley, New York.
5. Kerker, M. (1969) *The Scattering of Light and other Electro-magnetic Radiation.* Academic, New York.
6. Jackson, C., Nilsson, L. M., and Wyatt, P. J. (1989) Characterization of biopolymers using a multi-angle light scattering detector with size-exclusion chromatography. *J. Appl. Polym. Sci., Appl. Polym. Symp.* **43,** 99–114.
7. Burchard, W. (1992) Static and dynamic light scattering approaches to structure determination of biopolymers, in *Laser Light Scattering in Biochemistry* (Harding, S. E., Sattelle, D. B., and Bloomfield, V. A., eds.), Royal Society of Chemistry, Cambridge, UK.
8. Chapman, H. D., Chilvers, G. R., and Morris, V. J. (1987) Light scattering studies of solutions of the bacterial polysaccharide (XM6) elaborated by *Enterobacter* (NCIB 11870). *Carbohydr. Polym.* **7,** 449–460.
9. Garcia de la Torre, J. G. (1989) Hydrodynamic properties of macromolecular assemblies, in *Dynamic Properties of Biomolecular Assemblies* (Harding, S. E. and Rowe, A. J., eds.), Royal Society of Chemistry, Cambridge, UK.
10. Harding, S. E. (1987) A general method for modeling macromolecular shape in solution. *Biophys. J.* **51,** 673–680.
11. Harding, S. E. (1986) Applications of light scattering in microbiology. *Biotech. Appl. Biochem.* **8,** 489–509
12. Martinsen, A., Skjåk-Braek, G., Smidsrød, O., Zanetti, F., and Paoletti, S. (1991) Comparison of different methods for determination of molecular weight and molecular weight distribution of alginates. *Carbohydr. Polym.* **15,** 171–193.
13. Corona, A. and Rollings, J. E. (1988) Polysaccharide characterization by aqueous size exclusion chromatography and low angle light scattering. *Separation Size Technol.* **23,** 855–874.
14. Johnson, P. and McKenzie, G. H. (1977) A laser light scattering study of haemoglobin systems. *Proc. R. Soc.* **199B,** 263–278.
15. Wyatt, P. J. (1992) Combined differential light scattering with various liquid chromatography separation techniques, in *Laser Light Scattering in Biochemistry* (Harding, S. E., Sattelle D. B., and Bloomfield, V. A., eds.), Royal Society of Chemistry, Cambridge, UK, pp. 35–58.
16. Smidsrød, O. and Andresen, I.-L. (1979) *Biopolymerkjemi.* Tapir Press, Trondheim, p. 136. *See also* p. 309 of ref. 2.
17. Harding, S. E., Sattelle, D. B., and Bloomfield, V. A. (eds.) (1991) *Laser Light Scattering in Biochemistry.* Royal Society of Chemistry, Cambridge, UK.
18. Stacey, K. A. (1956) *Light Scattering in Physical Chemistry.* Academic, New York, pp. 86,87.
19. Stacey, K. A. (1956) *Light Scattering in Physical Chemistry.* Academic, New York, pp. 98–100.
20. Johnson, P. (1985) Light scattering–traditional and modern–in the study of colloidal systems. *J. Surf. Sci. Technol.* **1,** 73–85; *see also* Johnson P. and Ottewill, R. H. (1954) *Disc. Faraday. Soc.* **18,** 327–337.

[illegible]

[illegible] Jackson [illegible] and [illegible] (19[illegible]) [illegible]

[illegible]

[illegible] *Physical Chemistry* [illegible]

[illegible] *Sci.* **18**, 321–331.

CHAPTER 8

Determination of Diffusion Coefficients of Biological Macromolecules by Dynamic Light Scattering

Stephen E. Harding

1. Introduction

The importance of transport phenomena in biological processes is indisputable, whether they be concentration gradient driven, "active," or Brownian diffusion processes. Light scattering can provide a rapid probe into these processes, particularly if the technique of *dynamic* light scattering is used.

The technique of dynamic light scattering is possible because of the high *coherence* of laser light—that is to say, the light is emitted from the source as a continuous wave train. The wavelength of the otherwise highly monochromatic incident radiation can be Doppler broadened by the motion of the scattering particles: This broadening can be measured by a wave analyzer, and from the wavelength spread, diffusion coefficients can be measured. More commonly now, however, instead of a wavelength analyzer, the short-time fluctuations in *intensity*—caused by the movements of the macromolecules—are measured. These changes in intensity (or numbers of photons received by a detector) are recorded using a "Correlator," and from suitable analysis of the change of the "autocorrelation function" with time, translational and, in some cases, rotational diffusional information about the macromolecule can be obtained. Because of these features, "dynamic light scattering," like

From: *Methods in Molecular Biology, Vol. 22: Microscopy, Optical Spectroscopy, and Macroscopic Techniques* Edited by: C. Jones, B. Mulloy, and A. H. Thomas

classical light scattering (*see* Chapter 7), also comes with a plethora of alternative names: "intensity fluctuation spectroscopy" (IFS), "light beating spectroscopy," "photon correlation spectroscopy" (PCS), or "quasielastic light scattering" (QLS).

Other introductions to the technique can be found, for example, in van Holde *(1)* or from a short article by Johnson *(2)*. A detailed review of the methodology has been given by Bloomfield and Lim *(3)*. Applications to biochemistry can be found in, for example, refs. *4–6*. More complete mathematical texts are by Berne and Pecora *(7)* and Chu *(8)*. For a good description of how classical light scattering equipment can be modified for dynamic work, the reader is referred to Godfrey et al. *(9)*.

2. Summary of Information Obtainable

1. Translational diffusion coefficient, D (this will be a "z-average" if the system is heterogeneous).
2. The effective hydrodynamic or "Stokes" radius, r_H.
3. Gross conformational information in terms of bead modeling from diffusion coefficient data (in much the same way as the sedimentation coefficient is used—*see* Chapter 5).
4. An estimate for the polydispersity of a macromolecular solution (from, e.g., the "polydispersity factor," PF [normalized z-average variance of the diffusion coefficients]), or from various types of multiexponential inversion procedures.
5. Molecular weight (weight average), from the combination of D (z-average) with the (weight average) sedimentation coefficient.
6. An estimate for the rotational diffusion coefficient, D_R.

D and the corresponding r_H can be obtained relatively rapidly (a measurement can take <1 min in some cases), although D_R for asymmetric particles is much more difficult to obtain. D can be obtained to a precision of up to ~±0.2%; D_R to only ~±5% at best. The technique is particularly useful for looking at the time-course changes in size (and where appropriate, polydispersity) of assembling/disassembling system: for example, the kinetics of head-tail associations of T-type bacteriophages *(10)* or the effect of removal of Ca^{2+} ions on the swelling of southern bean mosaic virus *(11)*; another good example is the self-association of tubulin *(12)*.

3. Summary of Limitations

3.1. Sample Clarification: The "Dust" Problem

The smaller the macromolecule, the greater the problem: The technique is best suited for macromolecular assemblies (M $\gtrsim$ 100,000).

3.2. Asymmetry

Measurement at a single angle (conventionally 90°) gives insufficient information to obtain D for asymmetric scatterers—extrapolation to zero angle is necessary, which can cause problems, since at low angles, the dust problem is at its greatest.

3.3. Sedimentation

For very large particles (e.g., microbes), sedimentation under gravity can also contribute to the observed autocorrelation function.

4. Outline of Theory

Consider a scattering element in the fluid. Over short time intervals (~ns – μs), the positions and phase contributions of the particles within that element will fluctuate and, hence, the intensity of light scattered from that volume element will also fluctuate (Fig. 1). An "autocorrelator"—a purpose built computer—correlates intensities, $I(t)$, or equivalently numbers of photons $n(t)$, at time t with subsequent times $t + b\tau_s$, where b is the "channel number" (taking on all integral values between 1 and 64, or up to 128 or 256, depending on how expensive the correlator) and τ_s is a user-set sample time (typically 100 ns for a rapidly diffusing low-mol-wt [M ~ 20,000] enzyme, and increasing up to ~ ms for microbes). The product $b\tau_s$ is referred to as the "delay time" τ.

The correlator calculates the normalized intensity correlation function $g^{(2)}(\tau)$ as a function of the delay time τ:

$$g^{(2)}(\tau) = [\langle I(t) \cdot I(t + \tau)\rangle / \langle I \rangle^2] \tag{1}$$

The angular brackets indicate that the products are averaged over long times (user set as the "experiment duration time," which can be of the order of 1 min or higher depending on the size of the scattering particles and the power of the laser) compared with τ. As $\tau \rightarrow 0$, $g^{(2)}(\tau)$ can theoretically be as high as 2 *(13)* and decays with increasing τ to a lower limit of 1.

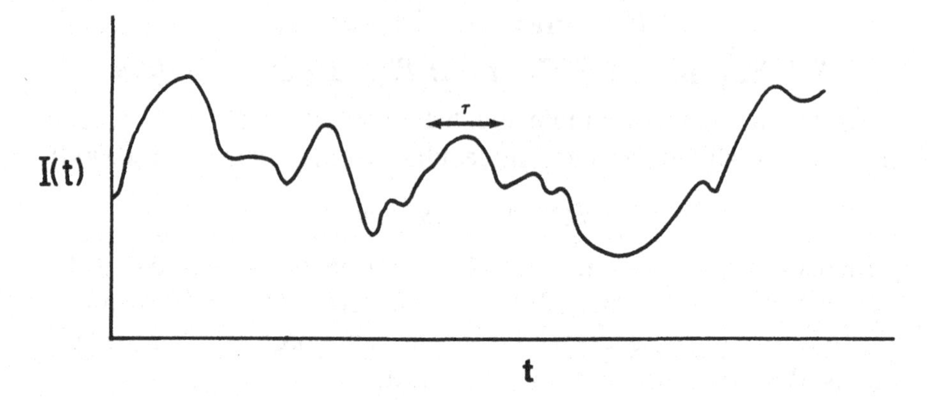

Fig. 1. Fluctuation in scattered intensity, I(t) with time, t. The variable τ is defined as the "delay time." At short delay times, there is good correlation in the scattered intensity; at long delay times, poor correlation.

For dilute Brownian systems (i.e., macromolecules and macromolecular assemblies with $M \leq 100 \times 10^6$), which are also quasispherical, the normalized autocorrelation function, $g^{(2)}(\tau)$ is related to the translational diffusional coefficient, D_2 by:

$$[g^{(2)}(\tau) - 1] = e^{-Dk^2\tau} \tag{2}$$

where k (or sometimes symbol "q" in the literature) is the Bragg wave vector whose magnitude is defined by:

$$k = \{4\pi n/\lambda\} \sin(\theta/2) \tag{3}$$

(n being the refractive index of the medium and λ the wavelength of the incident laser light). Thus, D can be found from a plot of $\mathrm{Ln}[g^{(2)}(\tau) - 1]$ vs τ, and an example is given for the motility protein dynein in Fig. 2.

The translational diffusion coefficient so obtained will be a function of solvent conditions, so as with the sedimentation coefficient (Chapter 5), it is usual to correct to standard conditions (water at 20.0°C), to give $D_{20,w}$ (*see*, e.g., ref. *1*). $D_{20,w}$ at a finite concentration will be an apparent value, and hence measurement at several concentrations and extrapolation to zero concentration to give an "infinite dilution" value, $D^{\circ}_{20,w}$, is normally necessary. However, the concentration dependence of $D_{20,w}$ is usually much smaller compared to other hydrodynamic parameters, such as $s_{20,w}$, and measurement at a single, dilute concentration may suffice.

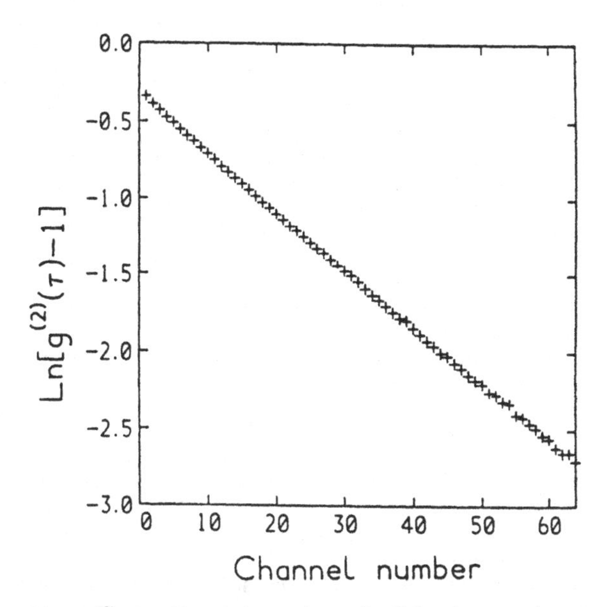

Fig. 2. Plot of Ln [$g^{(2)}(\tau) - 1$] vs "channel number" for the protein assembly dynein. τ is the delay time; $g^2(\tau)$ is the normalized autocorrelation function; "Channel number" = delay time/sample time.

4.1. Manipulation of $D°_{20,w}$

1. The $D°_{20,w}$ so obtained can be related to the equivalent hydrodynamic radius r_H of the macromolecule by Stokes equation:

$$r_H = (k_B T)/(6\pi\eta_{20,w} D°_{20,w}) \tag{4}$$

   where $\eta_{20,w}$ is the viscosity of water at 20.0°C.
2. Like $s°_{20,w}$, it can be used to obtain the frictional coefficient of the macromolecule and, from this, sophisticated "bead models" for macromolecular conformation (ref. *14*; *see also* Chapter 5).
3. $D°_{20,w}$ can be combined with $s°_{20,w}$ to yield an absolute value for the mol wt via the Svedberg equation *(15)*.

5. Experimental

A schematic dynamic light scattering setup is shown in Fig. 3. Collimated laser light from typically a helium-neon or argon ion laser light source is focused onto the center of a square (typically 1 cm²) or circular cuvet, placed at the center of a goniometer so that the scattering angle can be varied from typically 3° to 120°. Scattered light is

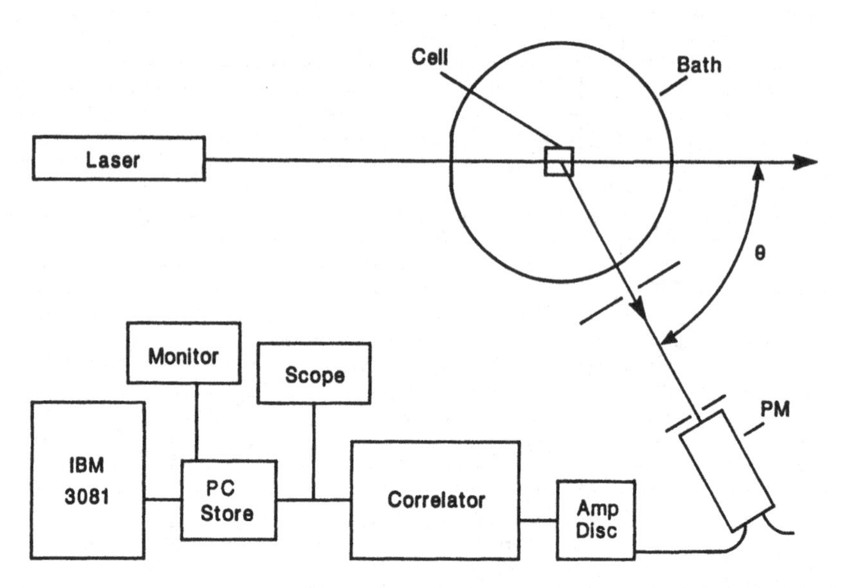

Fig. 3. Schematic dynamic light scattering apparatus (redrawn from ref. *16*). θ: scattering angle; Amp Disc: amplifier-discriminator; PM: photomultipliers; PC: IBM PC-compatible computer; IBM-3081: mainframe computer (only necessary for more sophisticated analyses).

collected by well collimated slits to reach the photomultipliers. Output from this is then processed via an amplifier-discriminator into a form suitable for processing in a digital correlator. Digital output from the correlator can be displayed directly onto an oscilloscope or via a microcomputer. In our laboratory, we store the basic correlation data on floppy disk via an IBM-PC compatible computer. The data can then be analyzed directly on the microcomputer or, for more detailed analysis, transferred to a mainframe computer. It is mandatory that the cell be kept in a thermostatted bath during measurement, since the diffusion coefficient is dependent on solvent viscosity, which itself is a strong function of temperature.

6. Commercial Availability of Instrumentation

Several dynamic light scattering photometers are available, and many of these are considered in ref. *5*. The principal manufacturers include Malvern Instruments and Biotage (UK), Coulter Electronics and Brookhaven Instruments (USA), Peters-ALV (Germany), and

Otsuka Electronics Ltd. (Japan). Most of the dynamic light scattering instrumentation also facilitates total intensity (i.e., "classical") light scattering measurements.

Although the software that comes with this instrumentation can considerably facilitate measurements, for the analysis of smaller macromolecular systems ($M \lesssim 100,000$), or for known asymmetric or self-associating macromolecules/assemblies, consultation with an expert user is recommended.

7. Preparation of Solutions

We would like to stress that the same attention to clarity of both sample and scattering cell that applies to the total intensity or "classical" light scattering method (*see* Chapter 7) is also necessary for dynamic light scattering. All solutions and cells need to be scrupulously free from even trace amounts of dust. We find that specially modified square-type fluorimeter cuvets (Fig. 4) are particularly useful for this purpose *(17)*. Repeated rinsing with ~ 0.2 μm filtered distilled water is necessary. If this procedure is not satisfactory, the use of acetone reflux apparatus may be necessary *(18)*.

Aqueous solvents should be of sufficient ionic strength to suppress charge effects. Loading concentrations required (which should be measured *after* clarification) will depend principally on (i) the size of the scatterer and (ii) the output of the laser. For example, if a 25-mW He-Ne laser is used, a loading concentration of at least ~1 mg/mL (and ~2–3 mL) is typically required, for a macromolecular assembly whose $M \sim 5 \times 10^6$. For macromolecules of mol wt down to ~10,000, more powerful lasers (~100 mW) and/or higher concentrations and/or longer experimental duration times are generally necessary to obtain meaningful results.

8. Notes

1. Choice of cuvet: Preferences vary, but we find that "square" cuvets are optically more reliable, since they do not suffer from total internal and stray reflection effects to the extent that can affect measurements using cylindrical cells. If measurement as a function of scattering angle is necessary, angles near the cell corners are obviously prohibited; if cylindrical cells have to be used, wide-diameter (i.e., ≥ 2 cm) cells are recommended.
2. Because of the enhanced effect of dust at low angles, a scattering angle of 90° is conventionally chosen for evaluation of D.

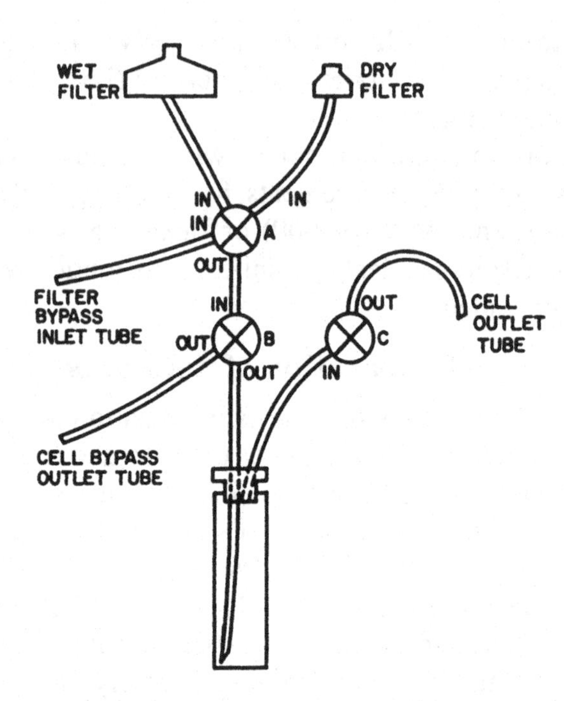

Fig. 4. Schematic cell filling apparatus for minimizing supramolecular contamination (reprinted with permission from ref. *17*).

3. Equation (2) is only exact for spherical particles. For nonspheroidal macromolecules scatterers, the contribution from rotational diffusional effects may not be negligible at higher angles, and the measured translational diffusion coefficient will be an apparent value with respect to angle. Therefore, in addition to measurement of $D_{20,w}$ as a function of concentration and extrapolation to zero concentration, a similar set of measurements as a function of angle and extrapolation to zero angle are necessary for asymmetric scatterers. These two extrapolations can be done on the same set of axes to give a "dynamic Zimm plot" (*19*). The form of the extrapolation to zero angle can, in principle, permit the determination of the rotational diffusion coefficient, D_R (*20*), although the precision with which D_R can be measured in this way is very limited. A good recent example of its careful measurement for a rod-shape virus is given in ref. *21*.
4. If the sample is polydisperse or self-associating, the logarithmic plot (Fig. 2) will tend to be curved, and the corresponding diffusion coefficient will be a *z*-average. The *z*-average $D^\circ_{20,w}$, when combined with

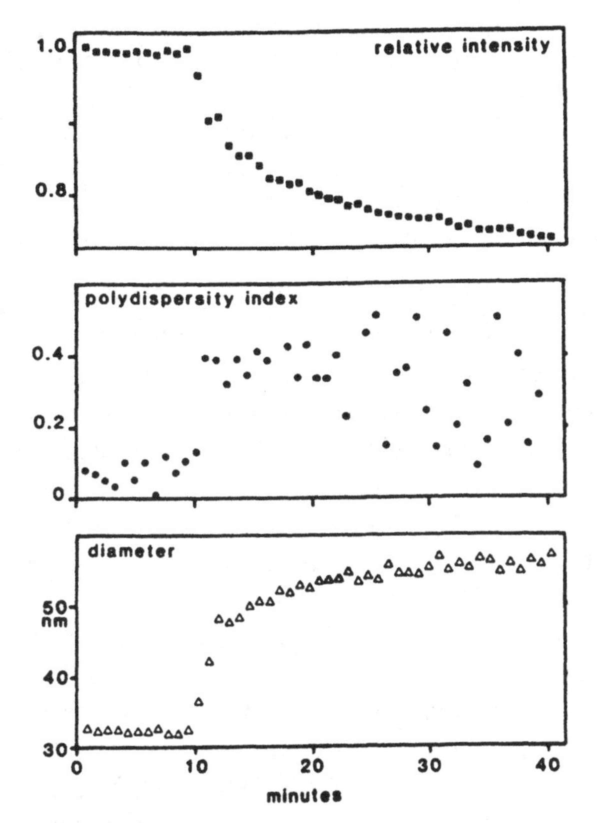

Fig. 5. Effect of removal of calcium ions (by adding EGTA at time $t = 10.5$ min) on southern bean mosaic virus. Top: total intensity (arbitrary units) scattered at 90°. Middle: polydispersity factor. Bottom: hydrodynamic diameter (from D). (redrawn from ref. *11*).

the (weight average) $s^{\circ}_{20,w}$ via the Svedberg equation, yields a *weight average* mol wt *(22)*, M_w.

5. Dynamic light scattering is particularly valuable for the investigation of *changes* in macromolecular systems. An example where it has been used to follow the swelling of a virus (southern bean mosaic virus) on removal of calcium ions is given in Fig. 5.
6. For a heterogeneous system, it is also possible as previously mentioned to obtain a parameter that indicates the spread of diffusion coefficients (the normalized *z*-average variance of the diffusion coefficients, referred to as the "polydispersity factor")—it is possible to relate this to the distribution of mol wt *(22)*. Commercial software, such as "CONTIN" *(23)*, is available for inverting the autocorrelation data directly to give

distributions of diffusion coefficient and equivalently particle size: These methods have recently been reviewed *(23,24)*. In addition, by analogy with LALLS/GPC and MALLS/GPC (Chapter 7), on-line coupling of dynamic light scattering to GPC has also been considered *(25)*.

7. For charged macromolecular systems, dynamic light scattering provides a useful tool for monitoring electrophoretic mobilities *(26)*. Commercial instrumentation is available for this purpose (*see*, e.g., ref. *5*).

Glossary of Symbols/Terms

QLS, Quasielastic light scattering; PCS, Photon correlation spectroscopy; IFS, Intensity fluctuation spectroscopy; D, Translational diffusion coefficient (cm^2/s) measured at a finite concentration; D_R, Rotational diffusion coefficient (s^{-1}) measured at a finite concentration; $D_{20,w}$, Translational diffusion coefficient at a finite concentration, c, and corrected to standard solvent conditions (i.e., water as solvent at a temperature of 20.0°C); $D^{\circ}_{20,w}$, Infinite dilution translational diffusion coefficient; k_B, Boltzmann constant (1.38062×10^{-16} erg/K); T, Temperature (K); $s^{\circ}_{20,w}$, Infinite dilution sedimentation coefficient (S or s); r_H, The effective hydrodynamic or "Stokes" radius of a particle (nm, µm, or cm); PF, Polydispersity factor (normalized z-average variance of the translational diffusion coefficients); M, Mol wt (g/mol); M_w, Weight average mol wt (g/mol); $n(t)$, Number of photons received by the photomultipliers at time t; b, Channel number; τ_s, Sample time (ns or µs); τ, Delay time (ns or µs) = $b\tau_s$; I, Intensity; $g^{(2)}(\tau)$, Normalized intensity autocorrelation function; k, Bragg wave vector (nm^{-1} or cm^{-1}); θ, Scattering angle; n, Refractive index of the medium; λ, Wavelength of the incident laser light (nm or cm).

References

1. Van Holde, K. E. (1985) *Physical Biochemistry.* Prentice Hall, Englewood Cliffs, NJ.
2. Johnson, P. (1984) Light scattering and correlation measurement. *Biochem. Soc. Trans.* **12,** 623–625.
3. Bloomfield, V. A. and Lim, T. K. (1978) Quasi-elastic light scattering. *Methods Enzymol.* **48F,** 415–494.
4. Sattelle, D. B., Lee, W. I., and Ware, B. R. (eds.) (1982) *Biomedical Applications of Laser Light Scattering.* Elsevier, Amsterdam.
5. Harding, S. E., Sattelle, D. B., and Bloomfield, V. A. (eds.) (1992) *Laser Light Scattering in Biochemistry.* Royal Society of Chemistry, Cambridge, UK.
6. Bloomfield, V. A. (1981) Quasi-elastic light scattering in biochemistry and biology. *Ann. Rev. Biophys. Bioeng.* **10,** 421–450.

7. Berne, B. J. and Pecora, R. (1976) *Dynamic Light Scattering: With Applications to Biology, Chemistry and Physics*. Wiley, New York.
8. Chu, B. (1974) *Laser Light Scattering*. Academic, New York.
9. Godfrey, R. E., Johnson, P., and Stanley, C. J. (1982) The intensity fluctuation spectroscopy method and its application to viruses and larger enzymes, *Biomedical Applications of Laser Light Scattering* (Sattelle, D. B., Lee, W. I., and Ware, B. R., eds.), Elsevier, Amsterdam, pp. 373–389.
10. Welch, J. B. and Bloomfield, V. A. (1978) Concentration-dependent isomerization of bacteriophage T2L. *Biopolymers* **17,** 2001–2014.
11. Brisco, M., Haniff, C., Hull, R., Wilson, T. M. A., and Sattelle, D. B. (1986) The kinetics of swelling of southern bean mosaic virus: a study using photon correlation spectroscopy. *Virology* **148,** 218–220.
12. Sattelle, D. B., Palmer, G. R., Clark, D. C., and Bayley, P. M. (1982) Oligomeric properties of brain microtubule protein characterized by quasi-elastic light scattering, *Biomedical Applications of Laser Light Scattering* (Sattelle, D. B., Lee, W. I., and Ware, B. R., eds.), Elsevier, Amsterdam, pp. 373–389.
13. Pusey, P. (1989) Dynamic light scattering by partially-fluctuating media, in *Dynamic Properties of Biomolecular Assemblies* (Harding, S. E. and Rowe, eds.), Royal Society of Chemistry, Cambridge, UK.
14. Garcia de la Torre, J. G. (1989) Hydrodynamic properties of macromolecular assemblies, *Dynamic Properties of Biomolecular Assemblies* (Harding, S. E. and Rowe, eds.), Royal Society of Chemistry, Cambridge, UK.
15. Tanford, C. (1961) *Physical Chemistry of Macromolecules*. Wiley, New York, p. 380
16. Johnson, P. (1985) Light scattering—traditional and modern—in the study of colloidal systems. *J. Surf. Sci. Technol.* **1,** 73–85.
17. Sanders, A. H. and Cannell, D. S. (1980) Techniques for light scattering from hemoglobin, in *Light Scattering in Liquids and Macromolecular Solutions* (Degiorgio, V., Corti, M., and Giglio, M., eds.), Plenum, New York, pp. 173–182.
18. Johnson, P. and McKenzie, G. H. (1977) A laser light scattering study of haemoglobin systems. *Proc. Roy Soc.* **B199,** 263–278.
19. Burchard, W. (1992) Static and dynamic light scattering approaches to structure determination of biopolymers, in *Laser Light Scattering in Biochemistry* (Harding, S. E., Sattelle, D. B., and Bloomfield, V. A., eds.), Royal Society of Chemistry, Cambridge, UK.
20. Aragon, S. R. and Pecora, R. (1977) Theory of dynamic light scattering from large, anisotropic particles. *J. Chem. Phys.* **66,** 2506–2516.
21. Johnson, P. and Brown, W. (1992) An investigation of rigid rod-like particles,in *Laser Light Scattering in Biochemistry* (Harding, S. E., Sattelle, D. B., and Bloomfield, V. A., eds.), Royal Society of Chemistry, Cambridge, UK.
22. Pusey, P. N. (1974) in *Photon Correlation and Light Beating Spectroscopy* (Cummins, H. Z. and Pike, E. R., eds.), Plenum, New York, p. 387.
23. Provencher, S. (1992) Low-bias macroscopic analysis of polydispersity in *Laser Light Scattering in Biochemistry* (Harding, S. E., Satteller, D. B., and Bloomfield, V. A., eds.), Royal Society of Chemistry, Cambridge, UK.

24. Johnsen, R. M. and Brown, W. (1992) An overview of current methods of analyzing QLS data, in *Laser Light Scattering in Biochemistry* (Harding, S. E., Sattelle, D. B., and Bloomfield, V. A., eds.) Royal Society of Chemistry, Cambridge, UK.
25. Rarity, J. G., Owens, P. C. M., Atkinson. T., Seabrook, R. N., and Carr, R. J. G. (1992) Light scattering studies of protein association, in *Laser Light Scattering in Biochemistry* (Harding, S. E., Sattelle, D. B., and Bloomfield, V. A., eds.), Royal Society of Chemistry, Cambridge, UK.
26. Ware, B. R. (1982) Membrane surface charges studied by laser Doppler electrophoretic light scattering, in *Biomedical Applications of Laser Light Scattering* (Sattelle, D. B., Lee, W. I., and Ware, B. R., eds.), Elsevier, Amsterdam, pp. 293–310.

PART III

Calorimetric Methods

CHAPTER 9

Introduction to Microcalorimetry and Biomolecular Energetics

Alan Cooper and Christopher M. Johnson

1. Introduction

As our knowledge of biomolecular structure becomes ever more detailed, it becomes increasingly important that we study the basic physical forces between and within macromolecules in sufficient detail, so that we might be able to understand and manipulate biological processes at the molecular level. Calorimetry is the only technique available for measuring these interactions directly, and microcalorimetric experiments have been performed on biological systems since at least the 1930s. It is only relatively recently, with improvements in calorimetric technology coupled with the advent of facile site-directed mutagenesis and genetic engineering methods, however, that these measurements have become relatively routine *(1–9)*. Commercial instruments are now available that are sufficiently sensitive, stable, user friendly, and cheap to allow microcalorimetry to become an almost routine analytical procedure in biochemical and biophysical research laboratories.

Microcalorimetry provides a means of studying directly the energetics of biomolecular processes at the molecular and cellular levels. Because heat effects are associated with almost all physicochemical processes, microcalorimetry may be used not only to determine absolute thermodynamic parameters for such processes as ligand binding, protonation, macromolecular assembly, conformational transition, and

From: *Methods in Molecular Biology, Vol. 22: Microscopy, Optical Spectroscopy, and Macroscopic Techniques* Edited by: C. Jones, B. Mulloy, and A. H. Thomas

so forth, but also as a general analytical tool to follow quite complex processes, such as cellular growth and metabolism, enzyme kinetics, and so on, over time scales ranging from a few seconds to several hours, and on samples that might be quite heterogeneous and turbid. Calorimetry falls into two broad classes depending on the kind of process under consideration. Differential scanning calorimetry (DSC) involves measurement of heat energy uptake (heat capacity) during changes of temperature, and is typically used for studying thermal unfolding processes of proteins and nucleic acids or phase transitions of lipid systems (*see* Chapter 10). By contrast, isothermal calorimeters are used for processes that occur or may be initiated at (essentially) constant temperature, such as substrate/inhibitor binding to proteins or small molecule intercalation into DNA (*see* Chapter 11); as well as more complex processes, such as light-induced reaction and even whole-organism metabolism. *See* refs. *1–11* for general references and review papers. Choice of instrumentation depends very much on the sort of process under investigation and how it might be initiated, and purpose-built instruments may be required. However, here we shall concentrate on the use of commercially available equipment in relatively straightforward studies of thermal stability, ligand binding, and related processes.

Although calorimetric techniques are absolute and measured heat effects are unambiguous (within the constraints of appropriate calibration, correction, and control experiments), interpretation of these heats requires some understanding of thermochemical principles and of the various experimental models. Moreover, the forces involved in stabilizing macromolecules and facilitating their interactions are overwhelmingly thermodynamic—for instance, they involve major entropic contributions that cannot be readily modeled by intermolecular potential functions or easily visualized in molecular models. Later chapters will cover some of the applications of calorimetry in more experimental detail, but the purpose of this first chapter is to summarize, in note form, some of the basic thermodynamic and thermochemical background common to all calorimetric methods. Much of this will be quite familiar to most potential users and may be found in standard texts, but is included here in an abbreviated form for convenience. Some other sections will be less familiar, and all parts should be read in conjunction with the appropriate experimental sections.

1.1. Energy Units

A small, but surprisingly contentious issue: Many authors continue to use the outmoded *calorie* instead of the accepted SI energy unit, the *joule*. Moreover, some instruments remain calibrated in calories. For the time being, at least, we have to live with both systems of units. Where necessary, energies will be quoted in both units, using the conversion factor: 1 cal = 4.184 J.

2. Basic Thermodynamics: Entropy vs Enthalpy

Calorimeters measure heats or enthalpies, but this is only one component of the forces governing any reaction. All equilibrium processes involving molecules are governed by free energy changes (ΔG), made up of both enthalpy (ΔH) and entropy (ΔS) contributions:

$$\Delta G = \Delta H - \mathrm{T} \cdot \Delta S \tag{1}$$

and processes will tend to go in the direction favoring reduction in free energy (ΔG negative). Free energy expresses the balance between the natural tendency for objects to attain as low an energy as possible (lower H, ΔH negative) and the equally natural, but usually opposing tendency for things to be as mixed up as possible (higher S, ΔS positive). Put another way: Since energy is conserved (First law of Thermodynamics), any decrease in energy in one part of a system must be at the expense of an increase elsewhere in the system (where the term "system" is normally taken to mean the object of interest plus its surroundings). What actually happens is governed by statistics at the molecular level, with its macroscopic manifestation as the Second law of Thermodynamics ($\Delta S_{\mathrm{TOT}} \geq 0$).

2.1. Forces/Interactions

This trade-off between the thermodynamic forces of entropy and enthalpy makes understanding of molecular interactions in solution difficult, since we need always to bear in mind the (usually invisible) changes taking place in the solvent. A familiar example is the hydrophobic interaction between nonpolar groups in water, where the bringing together of such groups is endothermic (ΔH positive), but is usually made favorable (ΔG negative) because of a larger contribution from a positive entropy change (ΔS positive). Both these effects arise from the solvent (water structure), so that the hydrophobic interaction, rather than being an attractive force between the groups in the usual sense, arises more from the mutual repulsion of solvent and solute components *(12,13)*. But even more familiar interactions can be dominated by solvent effects. Electrostatic forces between groups involve major solvation effects and may be endothermic or exothermic, depending on circumstances *(12)*. Even hydrogen bonding can be ambiguous. The formation of H-bonds between groups in proteins (during folding, for example) or between nucleic acids (base pairing), which we can see or infer from crystal structures, will usually involve the disruption of equally favorable, but usually invisible H-bonds with solvent water. Consequently, the overall contribution of hydrogen bonding to the ΔG of biomolecular processes in water might be quite small and not necessarily favorable *(14,15)*.

2.2. Heat Capacity

Both entropy and enthalpy are related to the heat capacity of the system. The enthalpy (H) may be visualized as the total energy (kinetic + potential: including rotations, translations, vibrations, interactions, and so forth) of all the molecules in the system, corrected for energy changes owing to PV work done during any volume changes. It may be imagined, and indeed measured, as the total energy required (at constant pressure) to heat the system from absolute zero to the required temperature, i.e.,

$$H = \int_0^T C_P(T) \cdot dT \qquad (2)$$

where $C_P(T)$ is the temperature-dependent heat capacity at constant pressure. (A similar relation holds at constant volume to give the internal energy U in terms of $C_V(T)$, but this is rarely used, since most biomolecular processes are measured under constant [i.e., atmospheric] pressure.)

The total entropy of any system may be expressed and measured similarly as:

$$S = \int_0^T [C_P(T)/T] \cdot dT \qquad (3)$$

Both these expressions for absolute entropy and enthalpy are equally valid for the more familiar *differences* in H and S:

$$\Delta H = \int_0^T [\Delta C_P(T) \cdot dT \qquad \Delta S = \int_0^T \Delta[C_P(T)]/T \cdot dT \tag{4}$$

where ΔC_P is the difference in heat capacity between the two states of the system. Over narrow temperature ranges, when ΔC_P stays roughly constant, the variation in enthalpy and entropy changes with temperature is given by:

$$\Delta H(T_1) \approx \Delta H(T_0) + \Delta C_P(T_1 - T_0) \tag{5}$$

$$\Delta S(T_1) \approx \Delta S(T_0) + \Delta C_P(T_1 - T_0)/T_0 \tag{6}$$

These expressions demonstrate the central role played by heat capacity in thermodynamics and explain the utility of DSC measurements, since particularly large changes in heat capacity occur during thermal transitions.

2.3. Standard States

All thermodynamic parameters will, in principle, depend on conditions, such as temperature, pressure, composition, and concentration. Consequently, it is necessary to adopt standard conditions ("standard states") for ease of comparison of data from different situations. Measurements are not necessarily always made under these conditions, but results are corrected to standard states, where feasible. Such data are usually denoted by the superscript 0 (ΔG^0, $\Delta C_P{}^0$, and so forth).

The standard pressure is normally taken to be 1 atm (760 mm Hg, 1.01325 bar, 1.01325×10^5 Pa). Standard temperature is usually 25°C (298.13 K), although sometimes 37°C is taken for biological systems.

For processes involving dilute solutions, the standard state for solutes is a concentration of 1*M*. (Strictly, this should be an activity of 1*M*, but activity coefficient corrections [which normally take account of nonspecific molecular interactions in solution] are rarely feasible in biomolecular experiments, nor are they justified by the available accuracy of the measurements.) An exception is normally made for pH-sensitive reactions in which H^+ is one of the active species. Here the standard state is chosen to be pH 7, with a more realistic hydrogen ion activity of $10^{-7}M$. Data using this convention are often indicated by a prime (e.g., $\Delta G'$, or $\Delta G^{0\prime}$).

A term that causes frequent misunderstanding is the standard free energy change (ΔG^0) for a process. This is the free energy change during the reaction when all species are present in their standard states. In particular, for processes in solution such as:

$$A + B + \ldots \rightarrow C + D + \ldots \tag{7}$$

ΔG^0 is the change in free energy that would be observed if all reactants and all products were present at $1M$ concentration (as well as standard T and P).

To obtain the free energy change (ΔG) under any other concentration conditions, we may use:

$$\Delta G = \Delta G^0 + RT \cdot lnQ \tag{8}$$

where T is the absolute temperature (Kelvin), R is the gas constant (1.987 cal $K^{-1}mol^{-1}$; 8.314 J $K^{-1}mol^{-1}$), and Q is the "reaction quotient"

$$Q = ([C][D]\ldots/[A][B]\ldots) \tag{9}$$

Part of the misunderstanding about ΔG^0 undoubtedly arises from the fact that, although it refers to free energy changes under standard concentration conditions, it can actually be measured from equilibrium concentrations, i.e.:

$$\Delta G^0 = -RT \cdot lnK \tag{10}$$

where K is the equilibrium constant for the process (i.e., the reaction quotient when the process has attained equilibrium). This relation forms the basis for most free energy determinations. Note that for any process that has reached equilibrium, $\Delta G = 0$ [but not $\Delta G^0 = 0$, except fortuitously for processes with $K = 1$].

2.4. *Van't Hoff Enthalpy* (ΔH_{VH})

A combination of the fundamental Eq. (1) and (10) shows how the equilibrium constant for a process is related to standard entropy and enthalpy changes, and to the absolute temperature:

$$lnK = -(\Delta H^0/RT) + (\Delta S^0/R) \tag{11}$$

Assuming that ΔH^0 and ΔS^0 do not vary significantly with temperature over the range of interest, differentiation gives the Van't Hoff equation (in one of its several manifestations):

$$\Delta H_{VH} = -R \cdot d\, lnK/d(1/T) = RT^2 \cdot d\, lnK/dT \tag{12}$$

Consequently, the enthalpy of a process may be estimated from (non-calorimetric) measurements of the equilibrium constant at different temperatures. The Van't Hoff plot (*lnK* vs 1/*T*) is frequently used, since it usually gives a convenient linear graph whose slope equals $-\Delta H_{VH}/R$. With very accurate data over a wide temperature range, the curvature of such plots may reflect the temperature dependence of ΔH and ΔS, and ΔC_P estimates might be made, but data are rarely reliable enough to justify this. One potentially useful feature of the Van't Hoff enthalpy is that it may be used, in conjunction with absolute calorimetric data, to determine the size of the reactive or cooperative unit taking part in the process of interest (*see* Section 4.1.).

3. Ligand Binding

Ligand binding may be studied by a variety of experimental techniques, but calorimetric methods (Chapter 11) offer the advantage of giving both equilibrium constant and enthalpy data simultaneously *(4,8)*. Although there are numerous pitfalls to avoid when analyzing data, particularly in systems with multiple binding sites *(16)*, here we shall merely summarize the simple single-binding-site situation.

Binding of a ligand molecule (L) to a single site on a macromolecule (M) may be described by the equilibrium equation:

$$M + L \leftrightharpoons ML \qquad K = [ML]/[M][L] \tag{13}$$

where the equilibrium constant, K (also known as the binding constant, association constant, or affinity constant; typical units: M^{-1}, mM^{-1}), is a measure of the strength of binding—larger K indicating tighter binding (more negative ΔG^0, *see* Eq. 10)

The dissociation constant (K_d = [M][L]/[ML]; typical units: M, mM, and so on) is the reciprocal of K and is a frequently used alternative. It has the advantage of indicating the free ligand concentration required to produce 50% saturation of binding sites.

When ligand and macromolecule solutions are mixed, heat energy is released (or absorbed) in proportion to the extent of binding and to the heat (enthalpy) of binding. The normal convention is that exothermic processes are denoted by a negative enthalpy, so the heat liberated per mole of macromolecule (Q), which may be measured directly by calorimetry, is given by:

$$Q = -[ML]\Delta H/C_M \quad (14)$$

where C_M is the total concentration of M (i.e., [M] + [ML]). Thus, as the ligand concentration is increased, the observed cumulative heat effect increases to a saturation value equal to the molar heat of binding per site. However, Q is also a measure of the proportion of sites occupied with a given ligand concentration, and may be used, just like any other probe of ligand binding (spectroscopic, and so on), to estimate K. The thermal titration curves may be analyzed by any of the standard procedures, but the most useful expressions, derived from the equilibrium equations above in terms of K_d, are the double-reciprocal form:

$$-\Delta H/Q = 1 + K_d/[L] \quad (15)$$

or the complete hyperbolic binding expression obtained using the substitution:

$$[L] = C_L + (Q \cdot C_M/\Delta H) \quad (16)$$

where C_L is the total ligand concentration. The simpler, double-reciprocal expression is more appropriate for analysis of weak binding situations where $[L] \approx C_L$ and a plot of $1/Q$ vs $1/C_L$ is approximately linear, with an intercept of $1/\Delta H$ and slope equal to $K_d/\Delta H$. However, this is at best, suitable only for rough preliminary estimates and in most cases the ligand binding affinity is sufficiently strong that it is no longer adequate to assume $[L] = C_L$. Full least-squares fitting of the data to the complete hyperbolic binding equation (or equivalent) is required, but this is normally within the compass of basic statistical software packages. Thus, calorimetric ligand-binding curves can yield estimates of both the enthalpy and equilibrium constant for the binding process, leading to ΔG^0 and ΔS^0 as well. Measurements at different temperatures will yield ΔC_P from the temperature dependence of ΔH. The temperature dependence of K, from the same experiments, will also give an estimate of the Van't Hoff enthalpy of binding (Eq. 12), which in comparison with the directly measured calorimetric enthalpy, can be used to estimate the number of (identical) binding sites on the macromolecule in simple cases (*see* Section 4.1.).

4. Thermal Transitions

Any equilibrium process brought about by increasing temperature must be endothermic, that is, heat energy must be absorbed in the process. Typical examples from the biomolecular field are thermal denaturation (unfolding) of proteins *(6,7,10,11)*, DNA/RNA "melting" or helix-coil transitions *(7,17–20)*, and the gel to liquid-crystalline phase transitions of lipid bilayers *(7,21)*. Information concerning the thermodynamics and cooperativity of such processes may be obtained using differential scanning microcalorimetry (DSC), which measures the excess apparent specific heat of a dilute sample with respect to a solvent baseline. (Exothermic processes are sometimes observed in DSC experiments, but these are almost always nonequilibrium effects brought about by enhanced rate processes at higher temperature.)

A typical DSC trace of a thermal transition (excess heat capacity vs temperature) will consist of one, or more, peaks indicating endothermic processes, with possibly a change in baseline (usually an increase) indicating a ΔC_P associated with the transition(s). Various levels of information may be extracted from such data.

The *midpoint temperature* (T_M) of the transition (usually, but not always the point of maximum C_P) gives a crude measure of the thermal stability of the system. For a simple two-state process, it is the temperature at which equal numbers of molecules will be in each state, or, taking a dynamic view, it is the temperature at which any molecule will spend exactly 50% of its time in each state.

The *integrated area* under the transition yields the absolute enthalpy change, Δh_{Cal} per gram (say) or ΔH_{Cal} per mole, for the process—*see* Eq. (4). Entropy changes may be similarly determined. Although in principle this analysis is independent of any model or assumptions about mechanism of the process, it does require care in selecting the appropriate baseline.

The *shape* of the transition depends on the cooperativity of the process. Highly cooperative processes, such as lipid-phase changes or macroscopic melting effects, give very sharp transitions, whereas less cooperative changes show much broader transitions. In appropriate circumstances, the apparent size of the cooperative unit involved in the transition may be estimated (*see* Section 4.1.).

Various procedures that might be used for detailed analysis of difference heat capacity data for macromolecular transitions and the deconvolution of overlapping transitions may be found in refs. *6,7,10,11*. Most methods rely on the so-called *two-state* approximation (*see* ref. *11* for some exceptions) in which the process is assumed to involve the temperature-induced shift between just two states, $N \rightleftharpoons D$ for instance in protein unfolding, with no significant accumulation of intermediate states (or, for multiple overlapping transitions, sets of separate two-state processes). The advantage of this model is that:

1. It conforms to conventional pictures of protein, nucleic acid, and lipid-phase transitions;
2. It is the assumption adopted in most noncalorimetric studies of conformational transitions; and
3. It can be expressed algebraically in relatively simple equations suitable for fitting DSC data.

In particular, within the two-state approximation, the accumulated heat energy uptake at any point within the transition (obtained by integrating the C_P curve) is a measure of the extent of reaction at that temperature. Consequently, it may be used like any other conformational probe (UV difference spectroscopy, CD, fluorescence, viscosity, and so forth) to estimate the equilibrium constant of the process as a function of temperature and, hence, using the Van't Hoff equation (Eq. 12), to determine the apparent enthalpy of the process. This Van't Hoff enthalpy (ΔH_{VH}) is independent of the calorimetric enthalpy (Δh_{Cal} or ΔH_{Cal}), and may be estimated even if the sample volume or concentration is unknown.

4.1. The Cooperative Unit: Δh_{Cal} *vs* ΔH_{VH}

DSC analysis of thermally induced transitions can yield two independent estimates of the enthalpy of the process. First, the directly measured total heat energy uptake in the sample during the transition, coupled with the known amount of sample in the calorimeter, gives the absolute enthalpy Δh_{Cal} per gram (say) of sample. Second, using the calorimetric signal as merely a probe of the extent of reaction at different temperatures and adopting the two-state approximation (or some similar model), yields the Van't Hoff enthalpy, ΔH_{VH} per mole. But this begs the question: "Per mole of what?" Answer: per mole of cooperative units *(1)*. Transitions may involve single molecules (monomers) acting individually. Alternatively, they may involve dimers, trimers, or much larger aggregates acting cooperatively. The size or apparent mol wt, m, of the cooperative unit may be estimated from the ratio of the Van't Hoff and calorimetric enthalpies for the transition:

$$m = \Delta H_{VH}/\Delta h_{Cal} \quad (17)$$

For most thermal unfolding transitions of small globular proteins, m is within 5% of the known monomer mol wt, and this has been taken as evidence in support of the two-state approximation for protein unfolding *(1,6,7,10)*. However, for much larger multidomain or multisubunit macromolecules, agreement is rarely so close. Broad transitions with low ΔH_{VH} and correspondingly low apparent m values indicate breakdown of the two-state approximation, with significant accumulation of intermediates or independent unfolding of separate domains. Van't Hoff analysis of highly cooperative lipid-phase transitions gives high apparent m values corresponding to cooperative units comprising several hundred lipid molecules *(1,18)*.

Similar considerations also apply to ligand binding and other experiments, involving comparison of Van't Hoff and calorimetric enthalpies. For the simple case of ligand binding to multiple identical binding sites on a macromolecule, m is the effective mol wt of each binding domain of the macromolecule. Comparison of this with the known size of the macromolecule (or oligomer) gives an indication of the number of binding sites.

5. Buffer Effects

5.1. Buffer Effects in Calorimetric Measurements

Calorimetric measurements using different buffer systems, under otherwise identical conditions, frequently give different heat effects. Sometimes this is because of buffer-specific effects on the system under investigation, but more often, it is a nonspecific effect arising out of proton (H^+ ion) uptake or release during the reaction. Consider the following generalized process involving the release of n protons:

$$X \rightarrow Y + nH^+; \quad \Delta H \quad (18)$$

In a well-buffered system, these H^+ ions, which would otherwise give rise to a pH change, are scavenged by buffer molecules (B) in a second reaction, which, although normally ignored, would give rise to an additional heat effect sensed by the calorimeter:

$$nB + nH^+ \rightarrow nBH^+; \quad -n\Delta H_I \tag{19}$$

where H_I is the molar enthalpy of ionization of the buffer species (*see* Table 1). Consequently, the observed enthalpy is a combination of the enthalpy of the reaction of interest (ΔH) plus the contribution from buffer protonation changes:

$$\Delta H_{Obs} = \Delta H - n\Delta H_I \tag{20}$$

Far from being an embarrassment, this calorimetric buffer effect is one of the most useful and accurate ways of measuring protonation changes, possibly unsuspected or hitherto undetected, in biomolecular processes (although it should be borne in mind that, following Le Chatelier's Principle or, more esoterically, the theory of linked functions *[27]*, any process that is pH dependent must involve protonation changes). Calorimetric measurements should routinely be performed using a range of buffers covering a range of ionization heats (Table 1) so that corrections may be made, where necessary. Ideally one should plot ΔH_{Obs} vs ΔH_I from a series of experiments under otherwise identical conditions (pH, temperature, ionic strength): the slope ($-n$) gives the number of protons released from the sample (or taken up, if n is negative) during the reaction, and the intercept (at $\Delta H_I = 0$) is the required heat of reaction, corrected for buffer effects *(28)*.

5.2. Buffer Effects in Van't Hoff Enthalpies

Nonspecific buffer effects are not confined solely to calorimetric measurements of enthalpies. The pH of most buffers changes with temperature (quite markedly in the case of amine buffers; *see* Table 1), and the heat of ionization of a buffer is directly related to this variation:

$$d(\text{pH})/dT = -\Delta H_I/2.303RT^2 \tag{21}$$

(Gas constant, $R = 1.987$ cal/K/mol[8.314 J/K/mol]; T in ^{0}K). Consequently, when studying any process that is sensitive to both temperature and pH (i.e., involves protonation changes), some care is necessary to discriminate between the two effects. This is particularly important when determining Van't Hoff enthalpies from the temperature dependence of equilibrium constants.

It is common laboratory practice to prepare the buffer of required pH at some convenient (room) temperature and then to use this buffer, unadjusted, for K measurements over the range of temperatures. In so

Table 1
Heats of Ionization and Related Data for Some Commonly Used Buffers[a]

		pK_A[b]	ΔH_I[c] kJ/mol	ΔH_I[c] kcal/mol	dpH/dT
Phosphoric acid	K_1	2.12	−7.9	−1.9	0.005
Glycine(–COOH)	K_1	2.35	4.2	1.0	−0.002
Citric acid	K_1	3.13	4.2	1.0	−0.002
Citric acid	K_2	4.76	2.5	0.6	−0.001
Acetic acid		4.76	0	0	0
Piperazine	K_1	5.7	29.3	7.0	−0.017
MES		6.1	14.6[d]	3.5[d]	−0.008
Cacodylic acid[e]		6.27	−2.0[f]	−0.47[f]	0.001
Citric acid	K_3	6.40	−3.3	−0.8	0.002
ACES		6.8	29.5[d]	7.05[d]	−0.017
Ethylene diamine	K_1	6.85	46.0	11.0	−0.027
PIPES		6.9	11.3	2.7	−0.007
Imidazole		6.99	36.6	8.75	−0.022
BES		7.19	24.3[g]	5.8[g]	−0.014
MOPS		7.2	20.5[d]	4.9[d]	−0.012
Phosphoric acid	K_2	7.2	3.3	0.8	−0.002
HEPES		7.5	20.5[d]	4.9[d]	−0.012
TES		7.55	32.1[g]	7.7[g]	−0.019
Triethanolamine		7.76	34.1	8.16	−0.02
EPPS (HEPPS)		8.0	20.9[d]	5.0[d]	−0.012
TRIS		8.0	47.44[h]	11.34[h]	−0.028
Diethanolamine		8.88	41.8	10.0	−0.025
Boric acid		9.2	13.8	3.3	−0.008
Ammonia		9.24	52.1	12.45	−0.031
Ethanolamine		9.5	50.6	12.1	−0.030
Piperazine	K_2	9.7	42.7	10.2	−0.025
Glycine(–NH_3^+)	K_2	9.78	44.4	10.6	−0.026
Ethylene diamine	K_2	9.92	49.8	11.9	−0.029
Phenol		10.0	23.4	5.6	−0.014
Phosphoric acid	K_3	12.4	12.5	3.0	−0.007
Water ($H_2O \rightarrow H^+ + OH^-$)		–	55.84[h]	13.35[h]	–

[a]Data taken from *(22)* unless otherwise indicated. 1 cal = 4.184 J.

[b]At 25°C, I ≤ 0.1*M*. Useful buffer range is usually $pK_A \pm 1$.

[c]For the process $HA \rightarrow H^+ + A^-$.

[d]Cooper and Nutley—unpublished observations, ±0.4 kJ/mol (±0.1 kcal/mol). Measured by HCl titration (Omega) of 10 m*M* solutions, pH adjusted to ca. pK_A, at 25°C.

[e]Not recommended for use in the presence of sulfhydryl agents *(23)*.

[f]Ref. *24*.

[g]Ref. *25*.

[h]Ref. *26*.

doing, one combines both the direct effect of temperature on K (i.e., ΔH_{VH}) with an indirect effect arising from the temperature-induced change in pH of the buffer (ΔH_I). Analysis of this situation (Cooper, unpublished) shows that apparent Van't Hoff enthalpies obtained in this fashion demonstrate exactly the same buffer dependence as shown above for calorimetric ΔH_{Obs}. Unfortunately, this effect is rarely recognized and even more rarely corrected for, and many published Van't Hoff enthalpies may be suspect on this account.

Methods for avoiding this problem in Van't Hoff analyses include:

1. Use of different buffer systems, as above;
2. Adjustment of buffer pH to the same value *at each temperature*; and
3. Choice of buffer system with $\Delta H_I \approx 0$ if convenient (e.g., acetate, phosphate).

5.3. Other Ligands (Coupled Reactions)

The calorimetric buffer effects described above for protons can be extended to other ligands. For instance, binding or release of Ca^{2+} or Mg^{2+} during protein transitions in the presence of EDTA or EGTA would give rise to additional heat effects associated with metal ion chelation. Similarly, calorimetric experiments on protein transitions involving exposure of -SH groups give much larger heat effects in the presence of sulfhydryl reagents (Cooper, unpublished). This emphasizes that calorimetric experiments monitor the totality of heat effects in any process. One must be aware of all the possible coupled reactions in the system and be alert to the possibility of exploiting them to advantage.

Acknowledgments

This work was supported by grants from the UK Science and Engineering Research Council.

References

1. Sturtevant, J. M. (1974) Some applications of calorimetry in biochemistry and biology. *Ann. Rev. Biophys. Bioeng.* **3,** 35–51.
2. Jones, M. N. (ed.) (1979) *Biochemical Thermodynamics*. Elsevier, Amsterdam.
3. Privalov, P. L. (1982) Stability of Proteins. Proteins which do not present a single cooperative system. *Adv. Protein Chem.* **35,** 1–104.
4. Wadsö, I. (1983) Biothermodynamics and calorimetric methods. *Pure Appl. Chem.* **55,** 515–528.
5. Ribeiro da Silva, M. A. V. (ed.) (1984) *Thermochemistry and its Applications to Chemical and Biochemical Systems*. NATO-ASI Series C, vol. 119, D. Reidel, Dordrecht.

6. Privalov, P. L. and Potekhin, S. A. (1986) Scanning microcalorimetry in studying temperature-induced changes in proteins. *Methods Enzymol.* **131,** 4–51.
7. Sturtevant, J. M. (1987) Biochemical applications of differential scanning calorimetry. *Ann. Rev. Phys. Chem.* **38,** 463–488.
8. Cooper, A. (1989) Microcalorimetry of protein-ligand interactions, in *The Enzyme Catalysis Process: Energetics, Mechanism and Dynamics* NATO-ASI Series A, vol. 178 (Cooper, A., Houben, J. L., and Chien, L. C., eds.), Plenum, New York, pp. 369–381.
9. Cooper, A. (1983) Calorimetric measurements of light-induced processes. *Methods Enzymol.* **88,** 667–673.
10. Privalov, P. L. and Khechinashvili, N. N. (1974) A thermodynamic approach to the problem of stabilization of globular protein structure: a calorimetric study. *J. Mol. Biol.* **86,** 665–684.
11. Freire, E. and Biltonen, R. L. (1978) Statistical mechanical deconvolution of thermal transitions in macromolecules. *Biopolymers* **17,** 463–479.
12. Kauzmann, W. (1959) Some factors in the interpretation of protein denaturation. *Adv. Protein Chem.* **14,** 1–63.
13. Tanford, C. (1980) *The Hydrophobic Effect.* Wiley, New York.
14. Klotz, I. M. and Franzen, J. S. (1962) Hydrogen bonds between model peptide groups in solution. *J. Am. Chem. Soc.* **84,** 3461–3466.
15. Tanford, C. (1970) Protein denaturation. *Adv. Protein Chem.* **24,** 1–95.
16. Klotz, I. M. (1989) Ligand-protein binding affinities in *Protein Function: A Practical Approach* (Creighton, T. E., ed.), IRL, Oxford, pp. 25–54.
17. Privalov, P. L. and Filimonov, V. V. (1978) Thermodynamic analysis of transfer RNA unfolding. *J. Mol. Biol.* **122,** 447–464.
18. Filimonov, V. V. and Privalov, P. L. (1978) Thermodynamics of base interaction in $(A)_n$ and $(A.U)_n$. *J. Mol. Biol.* **122,** 465–470.
19. Marky, L. A. and Breslauer, K. J. (1982) Calorimetric determination of base-stacking enthalpies in double-helical DNA molecules. *Biopolymers* **21,** 2185–2194.
20. Breslauer, K. J., Frank, R., Bloecker, H., and Marky, L. A. (1986) Predicting DNA duplex stability from the base sequence. *Proc. Natl. Acad. Sci. USA* **83,** 3746–3750.
21. McElhaney, R. N. (1982) The use of differential scanning calorimetry and differential thermal analysis in studies of model and biological membranes. *Chem. Phys. Lipids* **30,** 229–259.
22. Christensen, J. J., Hansen, L. D., and Izatt, R. M. (1976) *Handbook of Proton Ionization Heats.* Wiley, New York.
23. Cullen, W. R., McBride, B. C., and Reglinski, J. (1984) The reaction of methylarsenicals with thiols: some biological implications. *J. Inorg. Biochem.* **21,** 179–194.
24. Hu, C. Q. and Sturtevant, J. M. (1987) Thermodynamic study of yeast phosphoglycerate kinase. *Biochemistry* **26,** 178–182.
25. Vega, C. A. and Bates, R. G. (1976) Buffers for the physiological pH range:

thermodynamic constants of four substituted aminoethanesulfonic acids from 5 to 50°C. *Anal. Chem.* **48,** 1293–1296.
26. Grenthe, I., Ots, H., and Ginstrup, O. (1970) A calorimetric determination of the enthalpy of ionization of water and the enthalpy of protonation of THAM at 5, 20, 25, 35, and 50°C. *Acta Chem. Scand.* **24,** 1067–1080.
27. Wyman, J. (1964) Linked functions and reciprocal effects in hemoglobin: a second look. *Adv. Protein Chem.* **19,** 223–286.
28. Cooper, A., Dixon, S. F., and Tsuda, M. (1986) Photoenergetics of octopus rhodopsin: Convergent evolution of biological photon counters? *Eur. Biophys. J.* **13,** 195–201.

CHAPTER 10

Differential Scanning Calorimetry

Alan Cooper and Christopher M. Johnson

1. Introduction

A wide variety of temperature-induced transitions in biological systems may be studied by differential scanning calorimetry (DSC). This includes such processes as thermal unfolding (denaturation) of proteins, lipid membrane phase transitions, nucleic acid "melting," and so forth *(1–5)*. Such experiments may be analyzed to give not only the T_M and energetics of the transitions, but also information on the cooperativity/stoichiometry of the process *(6,7)*, as well as associated processes, such as protonation changes (*see* Chapter 9). A useful compilation of calorimetric and other data for protein transitions may be found in ref. *8*.

2. Instrumentation

Although a range of commercially available instruments exist, most of them are designed for small volumes (typically 50 µL) of solids or concentrated liquid samples, and are unsuitable for the majority of biophysical or biochemical studies involving dilute solutions or suspensions of biological macromolecules. Here we are concerned with situations where samples are typically 1 mg/mL or less, in 1–2 mL water or an aqueous buffer, where more than 99.9% of the heat capacity might arise from the solvent and must be accurately compensated for during the experiment.

Two DSCs designed specifically and widely used for such work are the DASM-4 (manufactured by the Bureau of Biological Instrumentation, Russian Academy of Sciences; contact VNESHBIO, Institute

From: *Methods in Molecular Biology, Vol. 22: Microscopy, Optical Spectroscopy, and Macroscopic Techniques* Edited by: C. Jones, B. Mulloy, and A. H. Thomas

of Bioorganic Chemistry, Miklukho-Maklai Street, 16/10, Moscow GSP-07, 117871 Russia) and the Microcal MC-2 (from Microcal Inc., 22 Industrial Drive East, Northampton, MA 01060). Both are very similar in basic principle and performance, being based on successful earlier designs by Privalov *(3,4)*, although the Microcal system currently offers the advantage of more sophisticated computer interface (PC compatible) and software for data collection and analysis. Microcal also supplies the Omega isothermal titration attachment to this instrument (*see* Chapter 11). Both systems now offer downscan options for experiments with decreasing temperature, as opposed to conventional increasing temperature scans, which may be useful for studying reversible folding/unfolding processes or cold-denaturation of proteins, for example *(9)*. Prices are fluid, but anticipate £40,000+ for a full DSC system (1992 prices, including PC-compatible microcomputer, plotter, and so forth). Required ancillary equipment will include a rotary vacuum pump for maintaining adiabatic chamber vacuum (Edwards E2M1.5, or equivalent, is satisfactory), N_2 cylinder and pressure regulator, and a refrigerated water bath/circulator (e.g., Techne RB-5) for jacket cooling and equilibration. A more sophisticated programmable circulating bath (e.g., Haake F3C) will be required with the downscan option.

The basic DSC comprises two matched cells, for sample and reference solutions mounted in an adiabatic chamber (Fig. 1). These cells are of a total-fill type and, as far as possible, are identical in volume and construction. Most satisfactory cells are usually pillbox shaped, with a loading volume of 1–2 mL, and are constructed of some suitably inert metal, such as tantalum, gold, platinum, and so forth. Other cell shapes have been used—for example, some versions of the DASM-4 have cells made from coiled capillary tubing for ease of filling, but these can cause problems because of the relatively large cell surface area in contact with the sample.

During operation in the conventional upscan mode, the temperature of the adiabatic shield is gradually raised by applying constant power to the main jacket heaters. Meanwhile, the feedback control systems monitor temperature differences between the cells and the jacket, and supply power to the cell heaters so that the cell temperatures follow the jacket as closely as possible. The difference in power supplied to the sample and reference cells is recorded as a function of temperature,

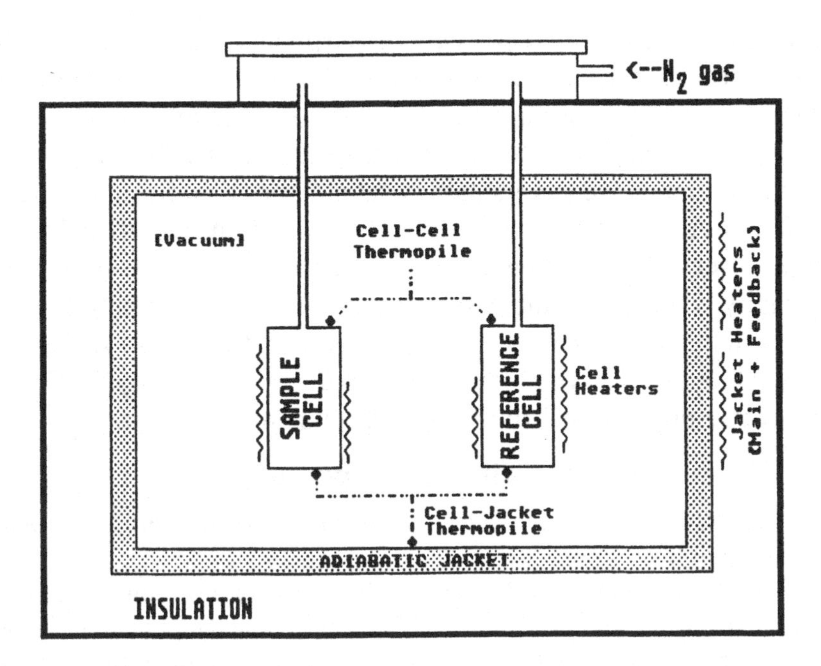

Fig. 1. Schematic view of the calorimetric unit of a typical differential scanning microcalorimeter. Sample and reference solutions, usually under inert gas pressure to inhibit bubbles, sit in identical, completely filled cells suspended by their filling capillaries in an evacuated adiabatic chamber. Sensitive thermopiles monitor cell–cell and cell–jacket temperature differences, while feedback heaters, controlled by external circuitry (not shown), maintain these differences constant and close to zero as the temperature of the system is raised steadily by main heaters on the adiabatic jacket walls. The differential energy input to the sample, compared to the reference, is recorded as a function of temperature (and time) to give the apparent excess heat capacity of the sample.

and is directly related to the heat capacity difference between them. The procedure in downscan mode is similar, except that the adiabatic jacket temperature is here reduced at a constant rate using a programmable refrigerated circulator.

3. Sample Requirements

Sample solutions should be equilibrated with the appropriate buffer (usually by dialysis) and the identical buffer used for reference and baseline scans. Solutions should normally be degassed before loading to avoid bubble formation at higher temperatures. The quantities of sample material required for DSC experiments will depend on the magnitude and sharpness of the transition(s) involved. For protein and nucleic acid unfolding transitions, solutions containing about 1 mg/mL (1.5–2 mL/scan) are usually necessary, although in favorable cases of small proteins, we have obtained satisfactory data from as little as 0.1–0.2 mg/mL. The much sharper phase transitions of purified lipids can be observed using even less material. Sample purity is important, especially when observing and attempting to interpret multiple transitions. Accurate estimation of sample concentration is also vital, since the absolute thermodynamic parameters derived from any DSC scan are directly proportional to the amount of sample present.

4. Running an Experiment

1. Equilibrate the sample with the appropriate buffer, preferably by overnight dialysis.
2. Degas both sample and buffer solution. This is best done in a small vacuum desiccator attached to a water aspirator and mounted on a magnetic stirrer. Removal of dissolved air in the sample (a particular problem with samples kept in the cold) can be seen by the appearance of small bubbles during gentle stirring under partial vacuum. Larger bubbles usually mean that the sample is boiling, and this should be avoided, since it will affect both sample and buffer concentrations and negate the equilibration. Two to three minutes of degassing is usually adequate and might even be omitted for delicate samples if it is not intended to scan to higher temperatures (>60°C) since thermal degassing is partially suppressed by the N_2 pressure head.
3. Measure the sample concentration (or retain an aliquot for later determination). UV/visible spectrophotometry is usually the most convenient method and, in many cases, may be done directly with the actual sample solution prior to loading. Bear in mind that the spectral properties of the sample may be altered during DSC if the transition is irreversible or aggregation occurs.
4. Load sample and buffer solutions into the appropriate DSC cells, avoiding entrapment of air bubbles. This is the trickiest bit. Microcal provides a suction-filling device for this, but we have found that manual filling is quite easy and less complicated. The usual procedure is to use a small syringe fitted with a long, blunt needle or (preferably) narrow-gage tubing (e.g., small animal cannula with Luer lock fitting)—long enough to reach the bottom of the DSC cell. Take up the appropriate

volume (1.5–2 mL) in the syringe and carefully displace any bubbles. Insert the syringe needle or tube into the cell inlet tube until the tip is just clear of the bottom of the cell, and inject the sample steadily until excess liquid just appears at the top of the inlet tube. Any trapped air bubbles can be removed by firm and rapid injection of the last 100–200 μL, sucking back and repeating several times if necessary. Wait a few seconds after each push to allow small bubbles to rise up the inlet tube. Repeat this procedure while withdrawing the syringe to clear the inlet tube of any trapped bubbles.

(Ideally, this filling procedure should be done starting with clean and dry cells. However, cell drying is time-consuming and can leave involatile solvent residues [smears] on the cell walls. We find it adequate to rinse the cells well with plenty of buffer before loading. However, since it is not possible to remove all drops of buffer from the cell, this can result in slight dilution of the sample on loading. This can be avoided with a prewash of the sample cell with a little of the sample solution [then discarded], if sufficient is available.)

5. Gently screw down the pressure seal cover on the DSC cells (finger tight), and apply nitrogen gas pressure. One to two atmospheres (15–30 psi) are adequate. The system should be sufficiently leak tight that both high- and low-pressure valves on the cylinder head may be closed without loss of pressure during the experiment.
6. Allow the system to equilibrate prior to starting the scan. The time involved will depend on various factors, including the temperatures of the sample and buffer when loaded and the temperature at which the scan is required to start, but for room temperature samples with a scan starting at about 20°C, this should not take longer than 10–20 min. Typically at the start of the active scan, one requires the temperature difference between cells and the adiabatic shield to be <1°C, with the temperature difference between sample and reference cells to be less than the equivalent of 0.5 mV on the first-stage preamplifier.

   (Subsequent procedures will depend on specific instrumentation and software [if any]. What follows is typical for ca. 1990 versions of Microcal instruments.)
7. Select scan parameters as required. Scan rate and temperature range are obvious, and will depend on sample requirements (60°C/h is satisfactory for many samples). "Filter" setting is slightly more obscure, but simply refers to the data acquisition and averaging time: e.g., a typical filter setting of 15 s will record and store averaged temperature and energy data every 15 s. Shorter filter times will give inherently more

noisy data and will rarely be required, except for very sharp transitions where enhanced resolution is needed. Be warned, however, that shorter filter times may result in a total number of data points over the scan that exceeds the software capability. (The equivalent adjustment for systems not under computer control usually consists of adjustment of the electrical time constant at the output stage to the X-Y recorder—usually a capacitor shunt on the output.)

Other facilities that may be selected at this time are options for repeat scans, with selection of equilibration/preequilibration times, standard upscan or (optional) downscan, and so forth. Switch the thermostat solenoid valve to automatic prior to start of computer-controlled scan. (If the feedback signals do not stabilize shortly after start of the scan, check: [1] that the thermostat solenoid valve is shut (i.e., not on manual override) and [2] that the thermal safety cutout is not triggered [press reset button].)

Representative data typical of a simple protein unfolding transition (using dilute lysozyme solution as example) are shown in Figs. 2 and 3. Note that, although the absolute values of the unnormalized data (Fig. 2) are arbitrary, the offset between buffer baseline and protein solution is real and illustrates that the heat capacity of protein (and, indeed, most organic substances) is lower than that of water, except during strongly endothermic transitions. The curvature of the raw baseline is also quite normal and reproducible, reflecting the small, but unavoidable differences in manufacture of the two cells.

8. At the end of the experiment, allow the cells to cool using the water bath, and remove the sample. (Samples may be removed while hot, but the cells should be filled with water immediately to prevent formation of dry sample film on the cell walls, which might be difficult to remove later.)
9. Before and/or after each experiment, identical scans should be performed using the appropriate buffer in both cells. Such buffer baselines are essential for correction of accurate quantitative data. (In a well-equilibrated and clean instrument, these baseline scans will be quite reproducible, but will show small differences with different buffer systems and will be markedly different at different scan rates.)

4.1. Analysis

Detailed analysis of collected data will depend on the type of sample and the model assumed for the process, but some basic operations should be performed that will be characteristic of all experiments:

1. Normalization to heat capacity: Current versions of Microcal DSC software collect differential thermal energy data as a function of time (mcal/

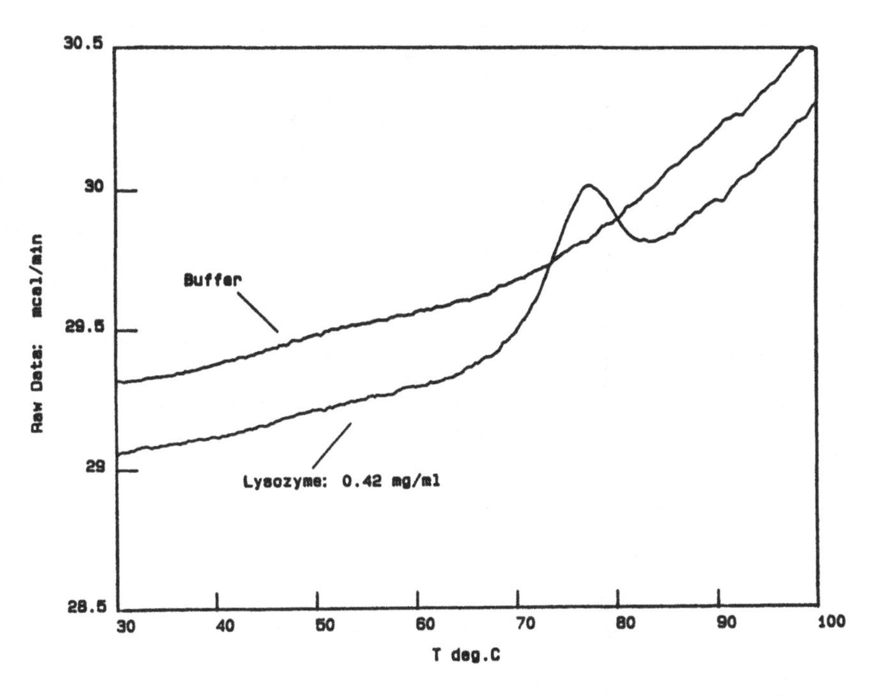

Fig. 2. Typical DSC (Microcal C-2) raw data for thermal denaturation of hen egg-white lysozyme, 0.42 mg/mL in 10 m*M* glycine/HCl buffer, pH 3.8., 60°C h^{-1} scan rate, 15 s filter, without smoothing, data normalization, or offset adjustment.

min)—i.e., a typical data file will consist of the time, cell temperature, and differential energy input to the cells since the last data point, at intervals determined by the filter setting. Conversion to differential heat capacity (mcal/degree) requires division by the mean scan rate at each point. Software options are available to do this automatically, but there is one (cosmetic) feature that is worth noting. Since measurements are always differential and buffer baselines are normally subtracted *(see* step 2 *below)*, the absolute value of the heat capacity signal (Y-value) is entirely arbitrary and can, indeed, be adjusted electronically on the instrument. However, because of variations in total heat capacity and heat losses in the instrument, the DSC scan rate varies slightly over the temperature range, and a large Y-axis offset can translate into a disconcerting curvature or deviation from the horizontal. This is an entirely cosmetic problem and disappears anyway when the similarly affected buffer baseline is subtracted, but it may be avoided by adjustment of the Y-axis offset control (use water in both cells) to give outputs close to zero (1–3 mcal/min). (Scan rate normalization is not normally required for systems not

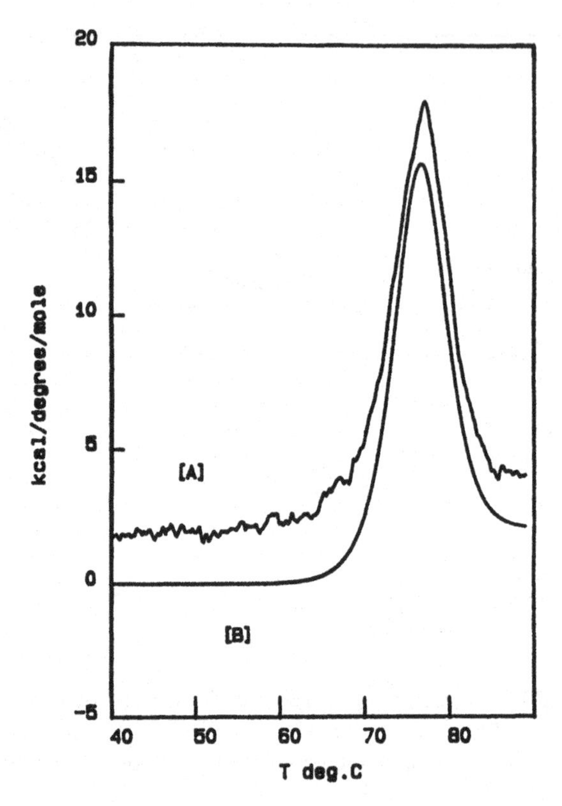

Fig. 3. [A] Data from Fig. 2 after normalization and baseline subtraction. [B] Theoretical best fit (offset for clarity) for a single cooperative transition with $T_M = 76.6°C$; $\Delta H_{Cal} = 119$ kcal/mol (498 kJ/mol); $\Delta C_P = +2.2$ kcal/K/mol (9.2 kJ/K/mol).

using computer control and data acquisition, since the X-Y recorder senses temperature directly [X-axis].)

2. Baseline correction: Subtraction of a (scan rate normalized) baseline will eliminate most of the artifact and nonlinearities arising from instrumental problems, such as slight mismatch of cell volumes, and so forth.
3. Concentration/volume normalization: Data for comparative purposes should be normalized to differential excess heat capacity per mole (say) by dividing by the actual amount of sample in the cell during the experiment (concentration × cell volume). Software options are available to do this, and the exact sample cell volume is determined by the manufacturer and supplied with the instrument.
4. The data are now ready for deconvolution in terms of the various models that might apply to the particular transition(s) involved *(4–7)*. Software is available with the MC-2 for analysis of the simpler situations (e.g., Fig. 3).

4.2. Calibration: Temperature

Standard samples are available for temperature calibration. These consist of small samples of pure hydrocarbons of known melting temperature sealed within steel capillaries that may be inserted into the DSC cell (filled with water). The very sharp melting transitions of these samples are readily observed and give a convenient check on temperature axis calibration (±0.1°C).

4.3. Calibration: Excess Heat Capacity

Accurate determination of transition enthalpies and heat capacities depends on combined knowledge of both the sample cell volume and the instrumental calibration (sensitivity). One is, to some extent, in the (usually reliable) hands of the manufacturer for these data, but it is useful, as a test of both instrumental performance and operator ability, to perform occasional independent calibration checks. Electrical calibration is normally provided, and this is satisfactory for routine checking, but it is by no means absolute (since one has to rely on the manufacturer's original adjustment and assume that it has not changed since) and, furthermore, does not test cell volume. In the absence of any convenient and sufficiently well-characterized thermal transitions, we have found that the differential heat capacity data for dilute NaCl and urea solutions are satisfactory *(10)*. The procedure is as follows.

With degassed water in both cell, record instrumental baselines over a 15–40°C range using normal instrumental settings. Prepare dilute solutions (1–2% accurately, by weight) of NaCl and/or urea in degassed water, and calculate the *molality* (mmol/kg of water).

Without changing any instrumental settings, load and scan these solutions as above, with water in the reference cell. The apparent heat capacity should be lower than baseline, and repeat scan (after refill) should be reproducible to within ±0.1 mcal/degree.

Normalize the data, divide by cell volume (mcal/min → mcal/degree/mL), and measure the decrease in heat capacity compared to pure water for each test solution at 24.15°C. This should be the differential volumetric heat capacity ($\Delta\Phi$) for which absolute values may be obtained from the data of Picker et al. *(10)*. Convenient expressions, obtained by regression analysis of these data, giving $\Delta\Phi$ in mcal/K/cm^3, are:

$$-\Delta\Phi(\text{NaCl},297.3\text{K}) = 35.62\,m - 10.55\,m^2 + 1.8\,m^3 \quad (1)$$

$$-\Delta\Phi(\text{Urea},297.3\text{K}) = 23.10\,m - 2.04\,m^2 \qquad (2)$$

where m = solute molality, 0–2 mol/ kg range. (Multiply by 4.184 for the equivalent values in mJ.) Measured values should normally be within 1% of these. Serious deviations (>5%) indicate instrumental problems or operator error.

5. Notes

1. Significant baseline fluctuations in DSC measurements may be caused by a poor or leaky vacuum system on the adiabatic shield. This is usually the result of poor connections or leaky taps. Reproducibility is best if the vacuum system is left running continuously.
2. If the instrument has been idle for some time, it might take a day or two to regain a reasonable standard of baseline reproducibility. Load the cells with degassed water (or buffer) and do several scans to high temperature (20–105°C say), overnight for convenience, until reproducible results appear. Similar overnight equilibration with buffer is advisable for most accurate work when changing to different buffer systems.
3. Experience will soon tell, from the instrumental parameters at the start of the scan, whether successful bubble-free filling has been accomplished. Remember that a 5-µL air bubble in a cell will produce a baseline offset of the order of 5 mcal/K (20 mJ/K), which is well outside the normal range of baseline variation. The direction of the offset with respect to the norm is a good guide to which cell contains the bubble: Positive offset indicates a bubble in the reference cell, and negative if in the sample.
4. Electrical interference: Occasional transient signals (spikes) may be caused by arcing from nearby electrical equipment: switching of refrigerators, air conditioning, heavy-duty motors, cheap thermostats, and so on, are frequent culprits. Stabilized power supplies or "spike filters" can help, but it is better to eliminate the source of interference. More regular interference ("grass") can sometimes arise from nearby microcomputers, which are inherently noisy devices.
5. Significant improvements in short-term noise (to $<\pm0.01$ mcal/min) may be achieved by use of an appropriate first-stage amplifier and good copper connections. The Model N2a DC nanovoltmeter from EM Electronics (Brockenhurst, Hampshire S042 7SJ, UK), or the more expensive Keithley instruments are ideal when used on the 1-mV input range (i.e., a preamplifier gain of 1000).
6. Temperature control: Ideally, any sensitive calorimetric equipment should be installed in a temperature-controlled environment. However, most room temperature controllers have an inevitable duty cycle that

although giving excellent long-term control, results in cyclic temperature fluctuations of the order of ±1°C with a period of order of 5–10 min. This can result in periodic baseline fluctuations. Our experience is that the DSC is best located away from drafts and heaters in a small room, preferably without windows, but certainly out of direct sunlight.

7. Reversibility/aggregation: Many samples, particularly high-mol-wt proteins, aggregate after unfolding and do not give reversible transitions. Typical symptoms of this are: asymmetric transitions, erratic and (sometimes) exothermic baselines after the transition, sensitivity to scan rates and concentration, and the appearance of turbidity in the sample. Repeat scan or downscan experiments on such samples are rarely satisfactory. Strictly speaking, thermodynamic analysis of such processes is not really valid, but it has been found *(4–6)* that, provided the irreversible steps are slow or delayed relative to the major unfolding transition, useful and usually consistent data may be obtained. Paradoxically, we have found that the effects of aggregation can sometimes be postponed by using slower scan rates (Microcal instrument). Although we do not understand this effect, it seems possible that irreversible aggregation of thermally denatured proteins may be accelerated by the enhanced convection currents present in the DSC sample cell at higher scan rates.
8. Take care when loading a fresh sample shortly after completion of a scan. Although the jacket temperature falls rapidly, the sample cell lags behind and can remain quite hot, threatening possible overheating of the next sample.
9. DSC cells should be cleaned regularly with some suitable detergent (e.g., Decon). Brief washing with 1–2% detergent followed by copious amounts of distilled water is sufficient for routine cleaning. When heavier contamination is suspected, detergent may be left in overnight with the instrument scanning to high temperature (up to 110°C if the N_2 pressure is on). Organic solvents may be used to remove lipids, and so forth.

Acknowledgments

This work was supported by grants from the UK Science and Engineering Research Council.

References

1. Sturtevant, J. M. (1974) Some applications of calorimetry in biochemistry and biology. *Ann. Rev. Biophys. Bioeng.* **3,** 35–51.
2. Privalov, P. L. and Khechinashvili, N. N. (1974) A thermodynamic approach to the problem of stabilization of globular protein structure: a calorimetric study. *J. Mol. Biol.* **86,** 665–684.

3. Privalov, P. L. (1980) Scanning microcalorimeters for studying macromolecules. *Pure Appl. Chem.* **52,** 479–497.
4. Privalov, P. L. and Potekhin, S. A. (1986) Scanning microcalorimetry in studying temperature-induced changes in proteins. *Methods Enzymol.* **131,** 4–51.
5. Sturtevant, J. M. (1987) Biochemical applications of differential scanning calorimetry. *Ann. Rev. Phys. Chem.* **38,** 463–488.
6. Privalov, P. L. (1982) Stability of proteins. Proteins which do not present a single cooperative system. *Adv. Protein Chem.* **35,** 1–104.
7. Freire, E. and Biltonen, R. L. (1978) Statistical mechanical deconvolution of thermal transitions in macromolecules. *Biopolymers* **17,** 463–479.
8. Pfeil, W. (1986) Unfolding of protein, in *Thermodynamic Data for Biochemistry and Biotechnology* (Hinz, H.-J., ed.), Springer-Verlag, Berlin, pp. 349–376.
9. Griko, Y. V., Venyaminov, S. Y., and Privalov, P. L. (1989) Heat and cold denaturation of phosphoglycerate kinase. *FEBS Lett.* **244,** 276–278.
10. Picker, P., Leduc, P.-A., Philip, P. R., and Desnoyers, J. E. (1971) Heat capacity of solutions by flow microcalorimetry. *J. Chem. Thermodynamics* **3,** 631–642.

CHAPTER 11

Isothermal Titration Microcalorimetry

Alan Cooper and Christopher M. Johnson

1. Introduction

Isothermal microcalorimetry is the generic term for a range of versatile and, in principle, nondestructive techniques suitable for direct measurement of the energetics of biological processes in samples ranging from homogeneous macromolecules up to complex and heterogeneous systems, including intact organisms *(1–6)*. We shall concentrate here on the use of calorimetry in the titration mode *(7–10)* to study the thermodynamics of ligand binding to macromolecules and related processes, since this is central to current efforts to understand the fundamental basis of macromolecular recognition processes and their applications in biotechnology.

It is worth emphasizing from the start that biomolecular binding experiments usually present much more of a calorimetric challenge than typical DSC measurements (*see* Chapter 10) and require considerably more sample material. This arises not only because noncovalent ligand-binding energies are intrinsically smaller than conformational transition enthalpies, for instance, but also because the total energy must usually be liberated gradually during progressive titration of a macromolecular binding site. Furthermore, ligand addition has additional heat effects (dilution, mixing) for which correction must be made, and which are frequently comparable to the binding heat effects.

Consider a possible estimate of the heat effects that one might wish to measure. Enthalpies of noncovalent binding of small molecules to protein sites are typically of the order 20 kJ/mol (5 kcal/mol; 1 cal =

From: *Methods in Molecular Biology, Vol. 22: Microscopy, Optical Spectroscopy, and Macroscopic Techniques* Edited by: C. Jones, B. Mulloy, and A. H. Thomas

4.184 J).* A 1- to 2-mL sample solution containing a few milligrams of protein (10^{-7} moles) would consequently liberate about 2 mJ (0.5 mcal) of heat energy during saturation binding. In practice, this would not normally be observed all at once, but rather as the accumulation of a series of heat effects during a thermal titration experiment involving successive ligand addition in which heat effects in any individual measurement might be 10% or less of the total. Measurement of this with an experimental accuracy of 10% or better would require instrumental sensitivity and noise levels down to 20 μJ (5 μcal) or less. This, corresponding to temperature changes in the sample of just a few millionths of a degree, is about the limit of current instrumentation, and is comparable to the inevitable heat effects of dilution, mixing, and stirring required in any binding experiment.

2. Instrumentation

Of the range of microcalorimeters available from various manufacturers, only those originating from LKB (now Thermometric) and, more recently, from Microcal Inc. have a reasonable track record for biophysical applications. These instruments are based on slightly different principles, which are worth describing separately.

2.1. LKB/Thermometric Thermal Activity Monitor (TAM)

Contact address: Thermometric AB, Spjutvägen 5A, S-17561 Järfälla, Sweden, or Thermometric Ltd., 10 Dalby Court, Gadbrook Business Park, Northwich, Cheshire, CW9 7TN, UK. The original LKB reaction microcalorimeters (batch and flow versions) still in use in many laboratories are based on designs by Wadsö *(7,9)* utilizing the so-called "heat leak principle." The Thermometric TAM is the direct successor based on the same principle but with updated electronics and improved temperature control (using a recycling water bath instead of air thermostat).

In outline, pairs of reaction vessels (cell) are contained within a relatively heavy and carefully temperature-controlled heat sink (Fig. 1). Between each vessel and the heat sink, and in good thermal contact with both, are solid-state thermopiles, which generate a voltage proportional to the rate of heat energy flux across them. The thermopiles

**See* the comment regarding energy units in Chapter 9.

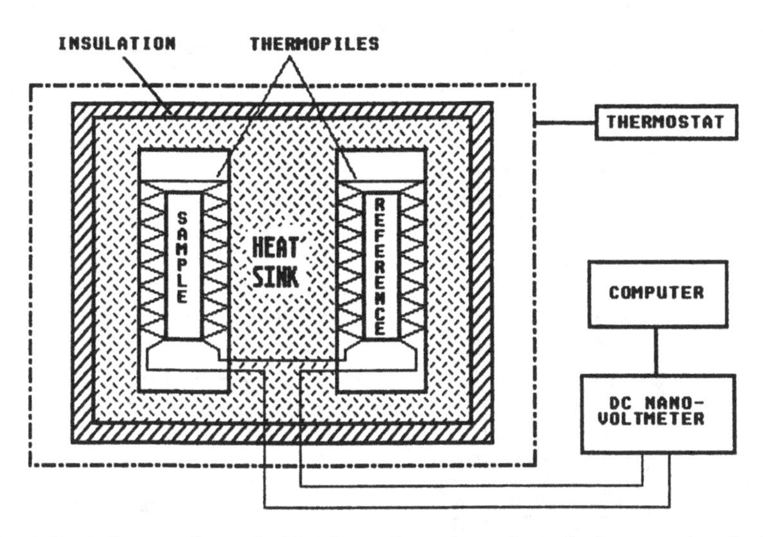

Fig. 1. Basic layout of a typical isothermal reaction microcalorimeter using the "heat-leak" principle. Sample and reference thermopiles are connected back-to-back in series to compensate for global temperature fluctuations and give differential measurements.

for each vessel are connected back-to-back in series for differential measurements, and one vessel would normally act as reference. For small temperature differences, ΔT, between any cell and the heat sink, both the heat flux across a thermopile and the potential (V) it generates are proportional to ΔT:

$$(dQ/dt) \approx k_1 \Delta T \text{ and } V \approx k_2 \Delta T \quad (1)$$

Hence

$$(dQ/dt) = k \cdot V \text{ and the heat of reaction } Q = k \int V \cdot dt \quad (2)$$

where k is the calibration constant. Integration over a heat pulse or measurement of the voltage offset during a steady-state experiment yields the desired heat.

This essentially passive measurement system has the advantage of simplicity and reliability, although these instruments tend to be rather slow to equilibrate and use. Simplicity of construction means that they are quite versatile and relatively easy to adapt for special purposes. A range of reaction vessels is available, or special cells may be constructed for different purposes (Fig. 2). Simple titration cells are usually fitted with motor-driven microliter syringes for delivery of ligand into fixed

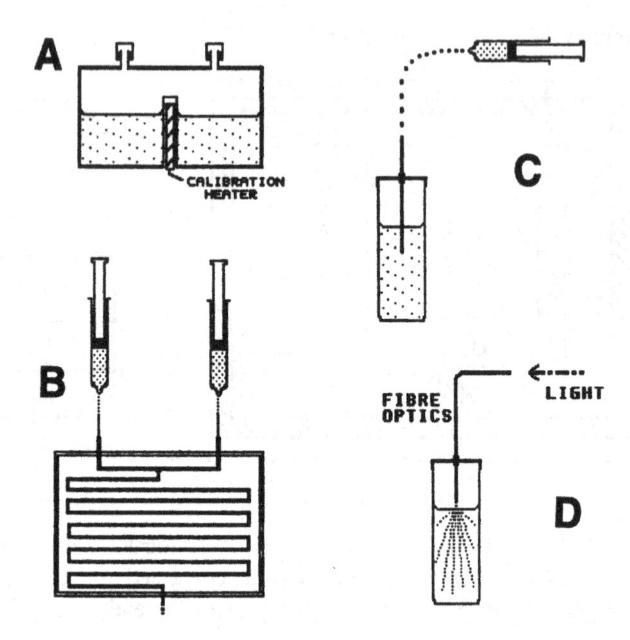

Fig. 2. Examples of cell configurations for different experiments in isothermal reaction microcalorimetry. All vessels will normally include electrical calibration heaters, shown here only in **A**. Vessels may be of glass or inert metals (Au, Pt, tantalum, Hastelloy™, and so on), and typical cell volumes are 1–10 mL. **(A)** Batch cell for single mixing experiments. Reactant solutions are loaded on either side of the dividing wall and, after thermal equilibration, are mixed by inversion/rotation of the entire calorimetric unit. **(B)** Flow cell for continuous or stopped-flow mixing experiments using either peristaltic pumps or, better, precision syringe pumps for delivery of samples. **(C)** Titration cell for *in situ* thermal titrations by microsyringe injection of ligand. Mixing is by inversion, as in **A**, or more recently using built-in stirrers. **(D)** Photocalorimeter cell for light-induced reactions. Samples are irradiated via flexible fiber optics light guides.

volumes of macromolecule solution. These syringes may be mounted in the thermostat or, alternatively, incorporate a heat exchanger coil in the delivery tube, so as to minimize spurious heat effects owing to lack of temperature equilibration of injected solutions.

2.2. Microcal Omega Rapid Titration Calorimeter

Contact address: Microcal Inc., 22 Industrial Drive East, Northampton, MA 01060. The Omega titration microcalorimeter is a more recent development utilizing a somewhat more complex active feedback system for determination of reaction heats *(8)*. It may be purchased as

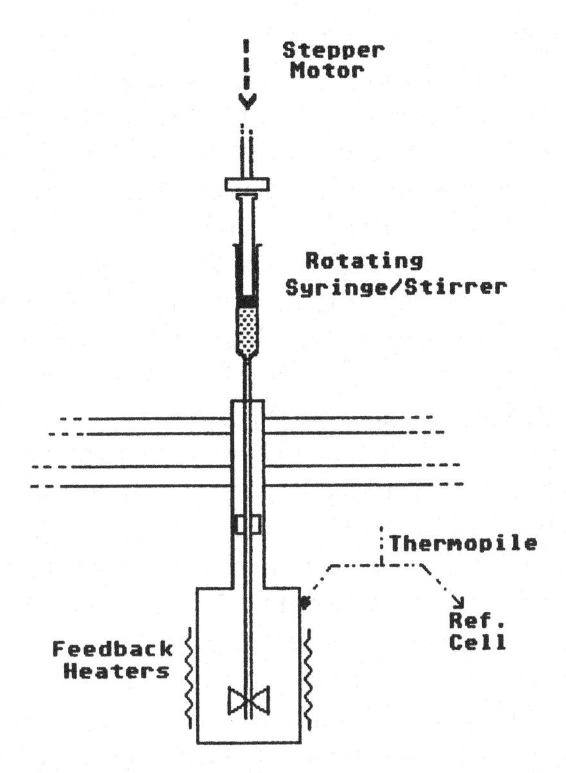

Fig. 3. Schematic of the Microcal Omega titration calorimeter cell. Each cell is suspended by its narrow filling tube in an evacuable adiabatic chamber. The reference cell (not shown) is connected to the sample cell by sensitive thermopile devices, and thermocouples also monitor the cell–jacket temperature difference. The delivery syringe, which is removed for filling and cleaning, is driven by a stepper motor drive and also rotates continuously during titration to provide stirring (the end of the needle is shaped as a narrow paddle).

a stand-alone instrument or as an attachment to the DSC instrument from the same manufacturer.

The reaction vessel consists of a simple cell fitted with a rotating syringe that acts as both delivery device and stirrer (Fig. 3). The reference cell is just a blank, unstirred and with no delivery system, usually kept filled with water. The principle of operation is very similar to the DSC from which it has been developed. Both cells are suspended on the ends of their filling tubes in an adiabatic chamber. (This chamber may be evacuated, although this is only usually necessary at low temperatures to prevent condensation.) Thermopiles mounted between both cells and between the cells and the adiabatic jacket measure temperature differences and, through a sensitive feedback system, control the feedback heater currents so as to maintain both cells and jacket at the same temperature. The difference in electrical energy required to maintain sample and reference cells at the same temperature is a measure of the energetics of processes occurring in the sample, and it is this energy that is recorded as a function of time during an experiment. As a consequence of the feedback process (the Omega is really a DSC running at very low scan rate), the temperature of the whole system tends to rise slowly with time, typically about 0.1°C/h or less, but the very slow temperature variation is rarely a significant problem.

Current versions of both instruments rely on microcomputer interfacing (PC-based) for instrumental control, data collection, and analysis. Software is usually provided for automatic integration of experimental peaks and for analysis of binding isotherms. In terms of sensitivity, there is little to choose between the Microcal and Thermometric systems. The Microcal Omega is significantly faster in operation: typically 2–3 min/injection, with a rapid turn around, so that complete titrations may be completed in an hour or less. The Thermometric TAM, on the other hand, requires several hours for each equilibration and titration sequence. However, it has the advantage of greater versatility over the Omega, with a wide range of cells available for different applications and good long-term stability.

3. Calorimetric Titration Method

Whichever instrumental system one uses, a typical calorimetric titration experiment consists of injection of a series of aliquots of ligand solution (with mixing) into a known volume of macromolecule solution. Each injection produces a heat pulse, which when corrected for ambient baseline and integrated over time, gives the total heat generated (or absorbed) in the event. There are four major sources of heat in such a process:

A. Macromolecule–ligand interaction (usually what we want);
B. Dilution of ligand on injection into the (normally) larger volume of sample;
C. Dilution of macromolecule by added ligand solution; and
D. Mechanical mixing effects.

In general, four separate measurements will be needed to discriminate them:

Titration	Heat Effects
1. Macromolecule/ligand	A + B + C + D
2. Macromolecule/buffer	C + D
3. Buffer/ligand	B + D
4. Buffer/buffer	D

Consequently, the required heat (A) for each addition of ligand is given by:

$$A = (1) - (2) - (3) + (4) \tag{3}$$

Sample preparation is usually quite straightforward provided that all solutions are equilibrated in exactly the same buffer, since even minor differences in buffer salt concentrations can give heats of mixing in excess of the ligand-binding heat required. The best procedure is normally to dialyze the macromolecule solution against a large volume of appropriate buffer, and to use aliquots of the final dialysis buffer to make up the ligand solution and for dilutions. Actual procedures and optimal concentration ranges will depend somewhat on the system under investigation *(8)*, but the following example will give guidance to the uninitiated.

3.1. Example: Lysozyme/N-Acetylglucosamine Binding

The binding of the simple monosaccharide, *N*-acetylglucosamine (NAG), to the specific site in the binding cleft of hen egg-white lysozyme (EC 3.2.1.17) provides a convenient and reliable test experiment for calorimetric titrations. This reaction has been well characterized *(11,12)*, and the materials are cheap and readily available. The binding is quite weak and exothermic with:

$$\Delta H = -5.81(\pm 0.2)\text{kcal/mol } (-24.3[\pm 0.8]\text{kJ/mol}) \tag{4}$$

and binding constant

$$K = 41(\pm 3)\ M^{-1}\ (K_d = 25 \pm 2\ \text{m}M) \tag{5}$$

3.1.1. Materials

Hen egg-white lysozyme (Sigma), *N*-acetylglucosamine (Sigma, St. Louis, MO). Buffer: 0.1*M* Na acetate adjusted to pH 5.0 with acetic acid (2–3 L).

3.1.2. Procedure

1. Prepare a lysozyme solution in buffer (about 10 mL, 5–10 mg/mL are adequate), and dialyze overnight in the cold against 2–3 L of buffer. Retain

this dialysis buffer for later dilutions. (Some protein samples may require centrifugation to remove insoluble material at this stage.)

2. Meanwhile, prepare the NAG solution, about 500 m*M* in the same buffer, and allow to stand for 3 h or more at room temperature to establish mutarotation equilibrium. (The α- and β-anomers of *N*-acetylglucosamine bind somewhat differently to lysozyme *[11]*.)
3. Prior to the experiment, briefly degas the lysozyme, NAG, and a sample of buffer solution. This is best done by placing the samples in small tubes or vials in a vacuum desiccator on a magnetic stirrer, evacuated with a water aspirator pump. Take care not to let the samples boil—just long enough for small air bubbles to be expressed.
4. Prepare the calorimeter for titrations at 25°C. Each titration will involve about 20 injections of NAG solution, about 5 μL each, into lysozyme or buffer solutions (1.5–2 mL).
5. Determine the concentration of the lysozyme solution by measuring the absorbance at 280 nm of accurately diluted aliquots (20–30-fold dilution normally required). A 1 mg/mL solution has an A_{280} of 2.65. Take the lysozyme mol wt to be 14,600.

   The following procedures are specific for the Omega titration calorimeter, since this is currently the instrument of choice for simple binding studies, but the description will serve as a useful guide for other systems. Bear in mind that some other systems are more versatile in that heats of dilution may be (partially) compensated directly by use of simultaneous titrations into a reference cell containing just buffer.
6. Totally fill the Omega sample cell with lysozyme solution, being careful to remove any bubbles that would produce noisy data. About 2 mL will be required. This is best done using a 2-mL disposable syringe fitted with tubing long enough to reach to the bottom of the cell (small animal cannula with Luer fitting is ideal). Load the syringe with solution, and express any bubbles. Insert the tube down the filling port of the calorimeter cell until it just touches the bottom. Inject the sample slowly, but firmly. The last few microliters should be injected rapidly to dislodge any bubbles, and this last step should be repeated several times.
7. Allow the sample cell to regain temperature equilibrium. This may take some time, especially if the sample is substantially cooler than the calorimeter. Equilibration may be accelerated by warming the sample to near the required temperature before loading—e.g., by using the heat of the hand around the loading syringe.
8. Meanwhile, totally fill the 100-μL titration syringe with NAG solution, avoiding bubbles. Rinse the syringe tip with buffer, and wipe dry.

9. Insert the full titration syringe into its rotating mount, and insert the assembled unit into the calorimeter. Care should be taken not to bend the needle or displace the plunger as the syringe tip is inserted into the sample. The unit should be pushed down firmly to form a good seal against the upper O-rings.
10. Swing the stepper motor and stirrer drive assembly back over the syringe, and attach the stirrer drive belt. Clamp the assembly down, and lower the stepper motor drive until it barely touches the syringe plunger. Attach the holding clip between drive piston and plunger. Switch on the stirrer motor, and adjust to about 400 rpm.
11. Subsequent thermal equilibration should be quite rapid (5 min). Use the software options to select an injection schedule of 20 × 5 µL injections, each of 30-s duration, 3–4 min between each. Start this injection schedule after a stable, noise-free baseline is obtained (50 µcal/s/200 µW range). Typical experimental results are shown in Fig. 4.
12. Repeat this procedure using the cell/syringe combinations: lysozyme/buffer (to give protein dilution heats), buffer/NAG (NAG dilution), and buffer/buffer (mixing effects).
13. Data analysis: Integrate each of the peaks using the software options provided. This is usually straightforward, but some care is necessary in the choice of appropriate baselines for noisy or drifting data. In the present example (Fig. 4), the exothermic lysozyme/NAG titration peaks range from 560 down to 270 µcal (2.34–1.13 mJ) during the titration (Fig. 4A). The NAG dilution heat (Fig. 4C), although roughly constant, decreases consistently from about 260 to 230 µcal (1.09–0.96 mJ) under identical conditions, and the best results are obtained by subtracting each peak from it corresponding titration partner. Additions of buffer to lysozyme solution or buffer alone (Fig. 4B) give identical, negligible heat effects in this case. The resulting thermogram (total heat effect vs total ligand concentration) may then be analyzed using standard least-squares procedures to give K and ΔH. For weak binding situations such as this, it is usually necessary to fix the number of binding sites ($n = 1$) to give sensible results, since the shallow curvature of the thermogram (Fig. 4) means that the stoichiometry of the binding is poorly defined.

3.2. Calibration

Electrical calibration involving standard heater resistors is provided with most instruments, and is quite accurate and satisfactory for routine calibration of thermal sensitivity. More complete calibration checks, which test the complete system including sample volume and

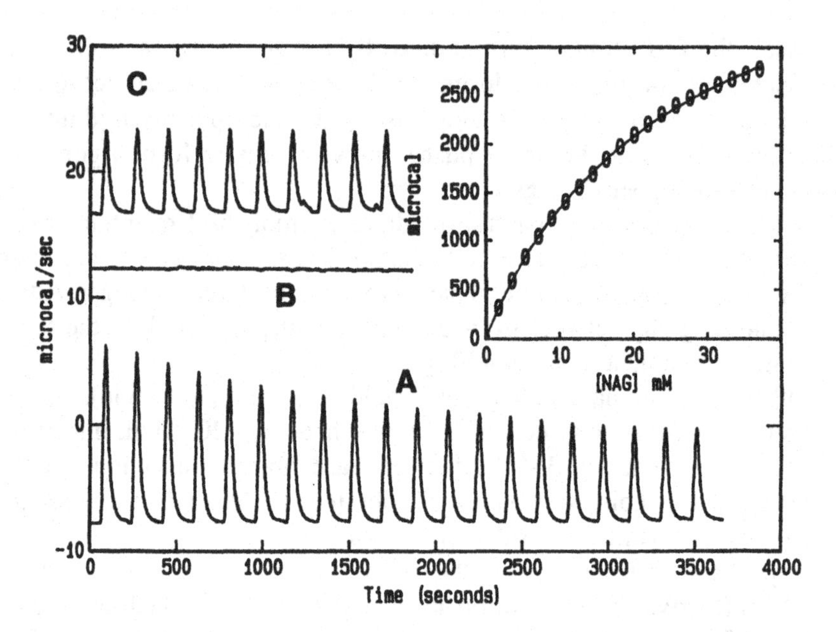

Fig. 4. Calorimetric titration of hen egg-white lysozyme (0.597 m*M*, 1.3963 mL) with *N*-acetylglucosamine (NAG), 0.1*M* acetate buffer, pH 5.0, 25°C using the Omega system. **(A)** 20 × 5 µL injections, 30-s duration, 3 min separation. **(B)** Lysozyme dilution, 10 × 5 µL injections of buffer into protein solution. **(C)** NAG dilution, 10 × 5 µL injections of 500 m*M* NAG into buffer solution. Insert: Integrated heat effects, corrected for dilution heats. The solid line is the least-squares fit for these data, calculated assuming a single binding site with $K = 40.2M^{-1}$, $\Delta H = -5.86$ kcal/mol (–24.5 kJ/mol).

titrant injection volumes, involve the use of standard reactions. The binding of *N*-acetylglucosamine to hen egg-white lysozyme (described in Section 3.1.) is now sufficiently well characterized *(11,12)* to act as a convenient test system. Simpler calibration reactions, giving larger heat effects, involve addition of standardized aqueous HCl to dilute Tris or NaOH solutions. For example, using the heat of ionization data at 25°C from ref. *13*, each 5-µL injection of 0.1*M* HCl into excess (10 m*M*) Tris base or NaOH solution should give exothermic heat effects of 23.7 mJ (5.67 mcal) and 27.9 mJ (6.67 mcal), respectively.

4. Additional Notes

1. Significant improvement in performance of earlier versions of both the Omega and LKB systems can be obtained by replacing the first-stage amplifier in the cell feedback circuit with a more stable and sensitive

system. Microcal originally recommended the Keithley Instruments Model 181 null detector, but a much more satisfactory (and considerably cheaper) amplifier is the Model N2a DC nanovoltmeter from EM Electronics (Brockenhurst, Hampshire SO42 7SJ, UK). Use of such an amplifier on the 1-mV range (i.e., preamplifier gain of 1000) reduces the RMS noise on the calorimetric signal (Omega, 2-s filter) to <0.005 µcal/ s. Fitting of the alternative amplifier is straightforward provided one takes care to avoid introducing spurious thermal emfs—use only pure copper connectors, *never* solder, and so on.

2. Room temperature fluctuations can cause baseline distortions in all kinds of microcalorimeters, either directly by affecting thermostat temperature and uniformity, or indirectly by producing spurious thermal emfs in the sensing circuitry. Place the instrument away from drafts and windows. Temperature control of the instrument room can help, but temperature fluctuations of up to ±1°C can occur during the normal duty cycle of most air-conditioner systems and can cause cyclic fluctuations in calorimetric baselines.
3. Electrical interference is a major problem, especially with the passive heat-leak type of instrument (LKB, and so forth), where the calorimetric signal is normally in the microvolt range. This usually shows up as spikes associated with switching of refrigerators, heaters, and other inherently noisy electrical equipment—even fluorescent lighting, on occasions. Microcomputers (PCs, and so on) can also be very noisy, usually producing continuous background interference ("grass" rather than "spikes"). Stabilized power supplies, "spike filters," screening of cables, and resiting of equipment can all help. Computer-based interference can sometimes be eliminated by switching the PC to another clock frequency.
4. Avoid β-mercaptoethanol in sample buffers. The slow air oxidation of this sulfhydryl reagent has a large associated heat effect that can cause large baseline offsets and drift. We have seen particularly large effects in calorimeters fitted with platinum reaction vessels, apparently because of Pt-catalyzed oxidation. Dithiols (DTT, DTE) present less of a problem since they are less prone to air oxidation.
5. Surprisingly large heat effects can arise from relatively minor microbial contamination of samples—the metabolic heat causing baseline offset and drift. Adopt clean working practices, and use Na azide in buffers if necessary. (**Caution:** Sodium azide is highly toxic. It also decomposes in acidic solutions liberating potentially harmful gas.)
6. Spurious and erratic behavior in continuous and stopped-flow reaction calorimeters can arise from temperature gradients in the system. This is a particular problem with older LKB-type instruments with air-bath ther-

mostats, where it is virtually impossible to achieve sufficient thermal stability and uniformity for sensitive binding experiments (although these instruments are ideal for less demanding experiments). Performance of these older instruments can be improved dramatically by mounting them in a water bath thermostat, ±0.001°C or better, with efficient stirring.

7. Evaporation from poorly sealed sample or reference cells (LKB batch type) can cause erratic behavior. This is not usually a problem with total-fill-type cells (Omega), where the liquid meniscus is at the top of the inlet tube and away from the thermally active part of the cell.
8. Long periods of equilibration with the injection syringe in place can lead to leakage or diffusion of ligand into the sample before the titration experiment starts. This effectively dilutes the first injection in the series and results in a lower than expected heat pulse. This can be minimized by delaying insertion of the loaded syringe until the sample cell has come almost to equilibrium after loading a new sample. Even so, trial experiments indicate that, in our hands, the first heat pulse of a titration series using the Omega 100-μL injection syringe might be up to 5% too low. Fortunately, the Omega software allows for thispossibility with an option to compensate for the effect. Similar problems in LKB and other titration systems may be minimized by use of very fine bore delivery tubing.
9. Heats of dilution depend on the change in concentration during the mixing process. This means that, in general, the dilution correction may be different for each injection in a titration series. The best procedure is to repeat each titration protocol with an identical injection sequence (omitting ligand or macromolecule, as appropriate) and to correct each binding heat pulse with its equivalent dilution partners in the sequence.
10. In experiments where binding is sufficiently tight that saturation is achieved during the titration (as indicated by the constancy of heat pulses from later injections), these later heat effects may be taken as an estimate of the combined dilution and mixing effects. This is recommended only as an approximate procedure, since it does not correct for possible concentration dependence during the titration (*see* Note 9).
11. Impurities: In theory, calorimetric experiments can tolerate any amount of contaminating material provided it does not interfere with the reaction of interest. This can be particularly useful with turbid samples, which might be difficult to study by spectroscopic methods. It is, however, necessary to know exactly the amount of active macromolecule in the sample, and unless unavoidable, extraneous impurities should be avoided. Also bear in mind that, at the relatively high protein concen-

trations usually required, even trace contamination with other enzymes (proteases, nucleases, phosphatases, and so on) may catalyze unwanted side reactions with ligand and produce spurious heat effects.

12. Isothermal microcalorimetry is in principle nondestructive, and samples may be recovered and recycled after use. However, some denaturation of proteins (for example) may occur during long periods of stirring and/or equilibration in the calorimeter, and samples should be tested for viability after the experiment.
13. Optimization of macromolecule and ligand concentration ranges for titration experiments is discussed in ref. *8*.

Acknowledgments

This work was supported by grants from the UK Science and Engineering Research Council.

References

1. Sturtevant, J. M. (1974) Some applications of calorimetry in biochemistry and biology. *Ann. Rev. Biophys. Bioeng.* **3,** 35–51.
2. Jones, M. N. (ed.) (1979) *Biochemical Thermodynamics.* Elsevier, Amsterdam.
3. Cooper, A. (1983) Calorimetric measurements of light-induced processes. *Methods Enzymol.* **88,** 667–673.
4. Wadsö, I. (1983) Biothermodynamics and calorimetric methods. *Pure Appl. Chem.* **55,** 515–528.
5. Ribeiro da Silva, M. A. V. (ed.) (1984) *Thermochemistry and its Applications to Chemical and Biochemical Systems.* NATO-ASI Series C, vol. 119, Reidel, Dordrecht.
6. Cooper, A. (1989) Microcalorimetry of protein-ligand interactions, in *The Enzyme Catalysis Process: Energetics, Mechanism and Dynamics* NATO-ASI Series A, vol. 178, (Cooper, A., Houben, J. L., and Chien, L. C., eds.), Plenum, New York, pp. 369–381.
7. Wadsö, I. (1968) Design and testing of a micro reaction calorimeter. *Acta Chem. Scand.* **22,** 927–937.
8. Wiseman, T., Williston, S., Brandts, J. F., and Lin, L.-N. (1989) Rapid measurement of binding constants and heats of binding using a new titration calorimeter. *Anal. Biochem.* **179,** 131–137.
9. Chen, A. and Wadsö, I. (1982) Simultaneous determination of ΔG, ΔH and ΔS by an automatic microcalorimetric titration technique. Application to protein ligand binding. *J. Biochem. Biophys. Meth.* **6,** 307–316.
10. Spokane, R. B. and Gill, S. J. (1981) Titration microcalorimetry using nanomolar quantities of reactants. *Rev. Sci. Instrum.* **52,** 1728–1733.
11. Cooper, A. (1974) Thermochemistry of binding of α- and β-*N*-acetylglucosamine

to hen egg-white lysozyme. Effects of specific oxidation of tryptophan-62. *Biochemistry* **13,** 2853–2856.

12. Bjurulf, C., Laynez, J., and Wadsö, I. (1970) Thermochemistry of lysozyme-inhibitor binding. *Eur. J. Biochem.* **14,** 47–52.
13. Grenthe, I., Ots, H., and Ginstrup, O. (1970) A calorimetric determination of the enthalpy of ionization of water and the enthalpy of protonation of THAM at 5, 20, 25, 35, and 50°C. *Acta Chem. Scand.* **24,** 1067–1080.

PART IV

Optical Spectroscopy

CHAPTER 12

Optical Spectroscopy

Principles and Instrumentation

Alex F. Drake

1. Introduction

Spectroscopy can be described as the study of the consequences of the interaction of electromagnetic radiation (light) with molecules. The most important interaction is absorption.

At the beginning of the 19th century, the corpuscular theory (particle theory) of light forwarded by Newton (1642–1727) was prevalent. However, this model was not capable of providing an adequate explanation for many of the optical effects being observed by this time. For example, Malus, investigating what we now recognize as the polarization of light on reflection, suggested that the particles of light were not hard; rather they were "soft" and capable of adopting an ellipsoidal rather than spherical form. Malus proposed that such an elongation occurs on reflection. By analogy with the N and S poles of a magnet, he called the reflected light "polarized."

The largely ignored philosophy of Huygens (1629–1695) that a wave theory of light was required to explain optical phenomena such as refraction and reflection was taken up by Young (1773–1829) and Fresnel (1788–1827) to explain and quantify observations that are now recognized as the result of the "interference" of one light wave with another. In the 1820s, Fresnel challenged the Newtonian view, not without reaction, and produced a mathematical model of a wave theory of light that explained most of the then known optical phenom-

From: *Methods in Molecular Biology, Vol. 22: Microscopy, Optical Spectroscopy, and Macroscopic Techniques* Edited by: C. Jones, B. Mulloy, and A. H. Thomas

ena. Central to his theme was the concept of linearly polarized light treated as a wave oscillating in a single plane. Unpolarized light implies that at all times all polarizations are equally likely.

Over 40 years later, in 1864, the ideas of Maxwell (1831–1879) in unifying the forces of electricity and magnetism led to the modern concept of electromagnetic radiation. The existence of propagating electromagnetic waves in a vacuum was confirmed experimentally in 1888, nine years after Maxwell's death, by Hertz (1857–1894), who invented oscillators to create the Maxwell waves and receivers to detect them. Electromagnetic radiation can be described as a propagating, oscillating electric field. Associated with any oscillating electric field there is an oscillating magnetic field. One is the consequence of the other, so only one need be explicitly considered. Spectroscopy involves the excitation of charged particles; thus, it is the electric field that is more relevant to the physical picture of a molecule sensing the light wave coming toward it.

A light beam has five important characteristics: wavelength (oscillation frequency), intensity, polarization, speed, and propagation direction (*see* Fig. 1 and Table 1). There is often confusion between energy and intensity. The energy of a light beam is related to its wavelength and dictates the type of interactions that can occur, whereas intensity is responsible for the number of these transitions excited. The intensity of the light beam determines the number of electrons ejected from the photocathode and the current from a detector.

2. Energy and the Wavelength of Light

2.1. Planck's Equation

As a major contribution to the evolution of Quantum Mechanics, Planck (1858–1947) derived the famous equation that carries his name. The Planck equation tells us that the energy associated with electromagnetic radiation is given as the product of the Planck constant and the frequency of the radiation

$$E = h\nu = (hc/\lambda) \tag{1}$$

that is, $E \propto 1/\lambda$ or $E \propto \tilde{\nu}$ (wave numbers, $\tilde{\nu} = 1/\lambda$). The shorter the wavelength of light, the greater the energy associated with it. The ramifications of this simple statement can be summarized as seven features:

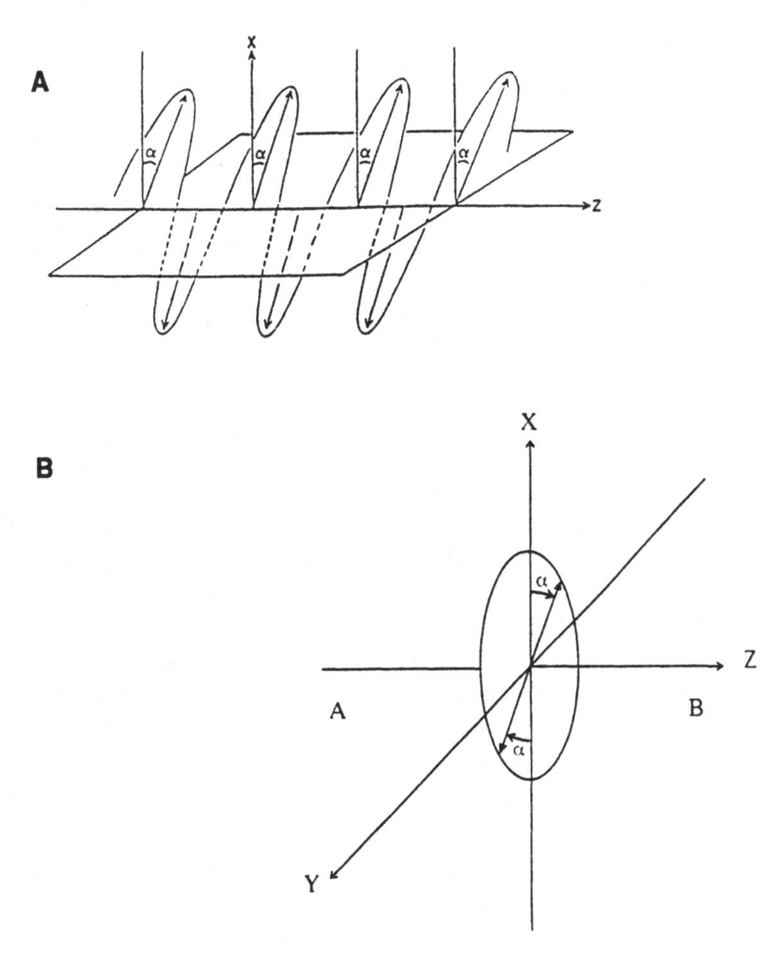

Fig. 1. The characteristics of light; where E is energy, h is Planck's constant, λ is wavelength, I is intensity, c is the velocity of light in a vacuum, and V is the velocity of light in a medium. **(A)** Space dependence picture. **(B)** Time dependence picture. The characteristics of light: (1) wavelength ($E = hc/\lambda$), (2) intensity (transmission I_0/I), (3) speed (refractive index $n = c/V$), (4) polarization, and (5) propagation direction.

1. Long wavelength radiowaves (and microwaves) are of a low energy that is sufficient only to "flip" the spin of nuclei (nuclear magnetic resonance, NMR) and electrons (electron spin resonance, ESR) (*see* Chapters 1 and 14, vol. 17 of this series). Spectral transitions are related to single, well-defined transitions between energy levels. Sharp spectral lines arising from a single transition may occur (possibly split by spin-spin coupling), if the lifetime of the excited state permits this. It is often possible to assign the spectral lines to particular atoms in the sample molecule.

Table 1
The Interaction of Light with Matter—Speed and Intensity

Total indices	Speed part, dispersion, real part	Intensity part, resonant absorption, imaginary part
Total absorption index	Refractive index (n)	Extinction coefficient (ε)
Total linear index	Linear birefringence ($n_{\|\|} - n_{\perp}$)	Linear dichroism ($\varepsilon_{\|\|} - \varepsilon_{\perp}$)
Total circular index (optical activity)	Circular birefringence (optical rotation) ($n_L - n_R$)	Circular dichroism ($\varepsilon_L - \varepsilon_R$)

2. Relatively higher energy microwaves of shorter wavelengths promote changes in the rotational and translational motions of molecules, which is the basis of microwave (rotational) spectroscopy.
3. Infrared radiation induces bonds to stretch, bend, and rock. Associated with each vibration energy is a series of sublevels distinguished by different rotational states, and a transition between rotational states may occur concurrently with the transition between vibrational states. Thus, each spectral line in an infrared spectrum arises from a family of overlapping, closely spaced transitions and is broadened (halfwidth about 10 cm^{-1}). Nevertheless, spectra are still composed of relatively sharp features, allowing the discrimination and assignment of vibrations associated with different bonds in a molecule.
4. Radiation of higher energy regions, the ultraviolet (UV) and the visible (Vis), will excite peripheral electrons away from the positive nucleus to a higher excited state. As an electron is excited, there will be associated changes in both the vibrational and rotational states of the molecule, and spectra become broadened. Readily excitable electrons are often delocalized about a group of atoms known collectively as a chromophore. (Beyond the UV, very high energy X-rays can cause ejection of electrons from very low-lying orbitals, and this is the principle of XANES and XAFS spectroscopy, described in Chapter 16, vol. 17 in this series). By tradition the nanometer unit is employed as the energy unit in the UV and Vis regions, accordingly bands in the visible often seem broader than those in the UV. Plotting in terms of wave numbers reveals that all electronic transitions have ideally a Gaussian bandshape with a similar bandwidth of about 5000 cm^{-1}, which corresponds to 20 nm at 200 nm (50,000 cm^{-1}) and 125 nm at 500 nm (20,000 cm^{-1}). This apparent variation in bandwidths is most noticeable in circular dichro-

ism (CD), where the signed character of the spectra inevitably leads to apparently sharper spectral components.

5. Sample holders and the environment will be at some temperature, typically room temperature, and the populations of the various energy levels will adopt a Boltzmann distribution. At a temperature, T, the number of molecules in an upper state (n_{upper}) relative to the number in a lower state (n_{lower}) is given by:

$$(n_{upper}/n_{lower}) = \exp\left[-(\Delta E/RT)\right] \quad (2)$$

A typical separation between energy levels in electronic spectroscopy is about 100 kJ/mol, IR vibrational spectroscopy is 10 kJ/mol, and in the radio region (nuclear reorientation) is 10^{-3} kJ/mol. Accordingly, the ratio (n_{upper}/n_{lower}) for electronic energy levels is 1.86×10^{-21}; for vibrational levels, it is 8.50×10^{-3}; and for electron and nuclear spin levels, it has a value on the order of 0.99999952. In other words, in electronic and IR vibrational spectroscopy, all molecules are effectively excited from their respective ground states. On the other hand, in NMR at equilibrium, nearly as many nuclei exist in the excited state as in the ground state, and when excited, the absorption of radiation is almost exactly balanced by that emitted as excited molecules drop down. The net absorption depends on the very small difference $n_{lower} - n_{upper}$, and this is a major limitation on the sensitivity of NMR.

6. In practice, an excitation is rarely "pure," since it does not take place in isolation and is sensitive to its molecular environment. In the infrared spectrum, the mechanical motions (vibrations) of different bonds can couple to give absorptions whose origin is complex. The ground and excited electronic states associated with substituted chromophores will differ from those of the unsubstituted chromophore. These factors can affect both the wavelength at which such transitions occur and their extinction coefficients.
7. The wavelength of electromagnetic radiation is much greater than the dimension of the average molecule. When the size of the absorbing species (molecular aggregates, whole ribosomes, liquid crystals, and so on) approaches that of the radiation, effects other than simple absorption take place (*inter alia* light scattering; *see* Chapters 7 and 8, of this volume.

2.2. Absorption Wavelength in the IR

In infrared spectroscopy, chemical bonds are often likened to springs with weights (atoms) at each end. The frequency at which this spring can be set to vibrate (absorb light) is governed by the strength of the spring (bond strength) and the mass of the weights (atomic weights), and the application of Hooke's Law is appropriate for a classical analysis. Thus, a typical IR spectrum is given below (Fig. 2).

Associated with any nonlinear molecule are (3*N*-6) possible fundamental modes of vibration, where *N* is the number of atoms in the molecule; the three translational modes (*x*, *y*, and *z*) and the three rotational modes do not give rise to vibrations. Macromolecules will therefore have hundreds of absorption peaks in the IR. Fortunately, these usually fall into distinguishable groups. The stretching of the O–H bond is perhaps the most difficult to set in motion and generally sets one limit of a typical IR spectrum at the relatively high energy of 3600 cm^{-1}. The fundamental N–H stretch is around 3400 cm^{-1}. The C–H stretches are around 2900 cm^{-1}. There is generally an energy gap until the region of the other important fundamental vibrations, which fall in the 1800–2000 cm^{-1} region. Perhaps the most prominent feature is the strong absorption associated with the C=O fundamental stretching vibration around 1600 cm^{-1} (*see* Chapter 14 of this volume).

The precise absorption wavelength is controlled by many factors. Thus, if the O–H group is hydrogen bonded (O–H---O) its absorption is shifted to lower wave numbers (lower energy) as the bond is weakened and lengthened. The very weak H---O "bond" has an associated stretching vibration at a very low energy, which is rarely detected, in the far IR. Another important effect is the coupling of vibrations. If in a molecule there is more than one vibration at a particular wavelength, then it is not difficult to imagine that these motions can couple to give "hybrid motions" with different spectroscopic properties. It is salutary to appreciate that in practice, most observed absorption maxima in the IR do not in fact correspond to pure, fundamental vibrations. The vibrations associated with the amide groups in polypeptides are not pure. Other features in the IR spectrum can correspond to overtones (two quanta of a particular vibration), combinations (absorption energy corresponds to the excitation of two vibrations of lower and individually different energies), and Fermi resonance (the interaction of an overtone with a fundamental).

The actual complexity of the IR spectrum means that it is more generally used as a fingerprint for structural identity. Overlapping contributions from different fingerprints are sought rather than an explanation of the precise molecular origin of the observed spectrum.

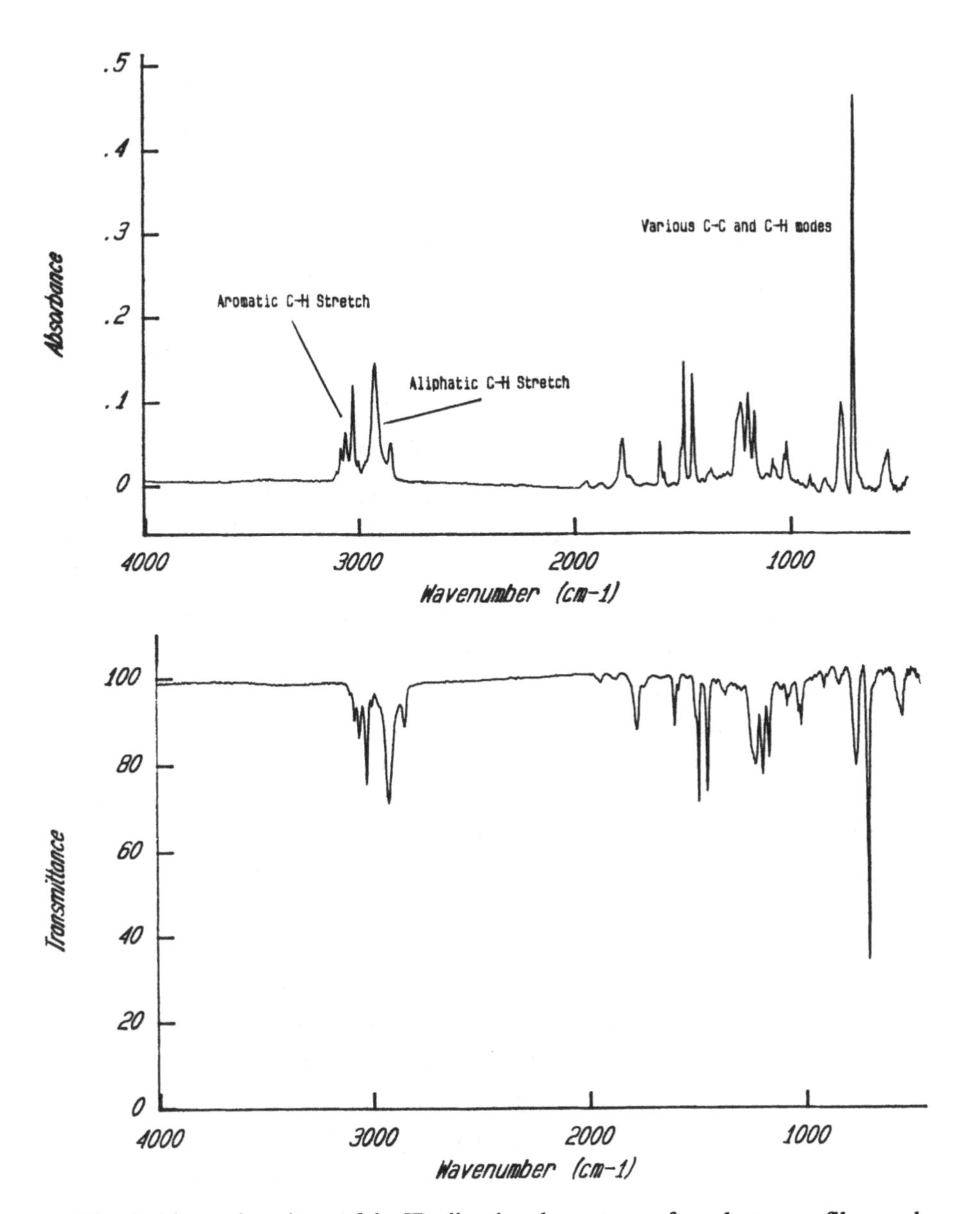

Fig. 2. Alternative views of the IR vibrational spectrum of a polystyrene film used for the calibration of wavelength.

In recent years, instrumental advances have meant that quantitative work is now more readily achieved. This is reflected in the trend to quote IR data in terms of absorbance (cf electronic spectroscopy) rather than transmission units (Fig. 2).

2.3. The Near IR Spectrum

The near IR runs from about 800 nm (12,500 cm^{-1}) to 4000 nm (2500 cm^{-1}). In this region, there are a few electronic transitions, largely those associated with extended π-systems, such as chlorophylls (around 800 nm) or weak metal ion based d-d excitations. Weak vibrational overtone and combination bands can be observed in the near IR. The first and second overtones, corresponding to a transition from the ground state not to the lowest energy excited state, but to a higher state, of the n_{OH} stretch of water at 1940 nm (5155 cm^{-1}) and 1430 nm (7000 cm^{-1}), respectively, are particularly useful for water determination.

2.4. Absorption Wavelength in Electronic Spectroscopy

Electronic spectroscopy is concerned with the excitation of an electron from the ground state orbital to an excited state orbital. The ease of excitation (wavelength) is dependent on how much higher in energy is the excited state level. The highest energy-bound electron associated with the C–C bond (σ bond) is only excited to a σ* antibonding orbital by very high-energy UV radiation at wavelengths below 150 nm. An electron from the lone pair on an oxygen atom requires wavelengths on the order 170 nm for excitation to a σ* state (*n*-σ transition). For organic molecules, absorptions occur in the UV only for unsaturated systems or polarizable substituents, like sulfur. The $\pi \rightarrow \pi^*$ transition of an isolated double bond (e.g., in cholesterol) occurs around 180–195 nm. Extension of the π-system, cyclically to benzene, naphthalene, purine, and pyrimidine or linearly to polyenes, such as carotenes, means that the highest energy π-electron is delocalized and more readily excited to one of several π* excited states. Eventually, with carotenes and complex π-systems, like chlorophylls, the excitation occurs with the relatively lower energy of visible and near IR light. If there is more than one chromophore in a molecule, they can interact in a manner akin to interacting IR vibrations, or an electron from one chromophore can be excited to a state localized on another (charge transfer).

In the carboxyl chromophore, the only valence excitation that has been assigned is the rotation of charge associated with the movement of an oxygen-based *n*-electron to a π*-excited state of the C=O bond occurring around 300 nm. The same is true of the ester or carboxylic acid groups around 210 nm. The amide group is claimed to have an *n*-π* transition around 200 nm with a $\pi \rightarrow \pi^*$ transition around 190 nm.

Transition metal ions have readily excited d-electrons giving d-d transitions that generally occur in the visible region (hence their color). The transfer of charge on excitation between the metal center and the ligand will occur in the UV as charge transfer transitions.

3. The Ability to Absorb Light

3.1. The Molar Extinction Coefficient

Energy levels can be traversed on resonant interaction with radiation of the appropriate wavelength. However, this does not imply that just because the right wavelength is available, there is a 100% probability that excitation will occur. Imagine you are confronted by a range of mountains with many peaks of the same height. Some of these peaks will be easy to get to by well-trodden paths, whereas others may be all but inaccessible. So for excitations, unless there is an allowed path to the excited state, an excitation will have a low probability of taking place even though energy of the correct wavelength is passing by. In measurement terms, this means that the molar extinction coefficient can be large for allowed transitions, taking values >6000. Forbidden transitions will have ε values <1000; severely forbidden transitions have ε values <1.

The ability of a molecule to absorb light is governed by selection rules. These rules are quantified in Beer's Law:

$$\log(I_0/I_t) = A = \varepsilon[c]l \qquad (3)$$

where I_0 is the intensity of the incoming irradiation and I_t is the intensity of the irradiation after passing through the sample; A is absorbance; ε is the extinction coefficient; $[c]$ is concentration; l is the path length of the sample. The magnitude of ε is a measure of the probability of the transition occurring (the ability of a molecule to absorb light).

Thus, toluene, typical of phenyl compounds, has three major absorptions associated with the phenyl ring with the values: $\varepsilon_{185} = 55{,}000$; $\varepsilon_{210} = 8000$; $\varepsilon_{260} = 200$. The two transitions of lower energy are formally forbidden, although the 210 nm excitation gains intensity by virtue of its proximity to the fully allowed 185-nm excitation.

3.2. Selection Rules

For a transition to be allowed, the excitation must be both symmetry and spin allowed.

3.2.1. Symmetry Selection Rules

Electromagnetic radiation is effectively an oscillating electric field in a plane transverse to the propagation direction. This will excite a charged particle most effectively if the excitation involves a translation of charge. This translation of charge gives rise to a transition electric dipole during excitation, and symmetry rules state that such a condition must pertain for an absorbance band to be allowed. Thus, in the illustrated example (Fig. 3), linearly polarized light polarized in the zy plane (as defined in Fig. 1) induces the movement of an electron (charge) along the C=C bond (oriented in the z axis) out of the π orbital into the π^* orbital. This movement results in an electric transition dipole moment along the z axis. The absolute direction of the moment is arbitrary, since the electron can be excited up or down the double bond as drawn with equal facility. This must not be confused with the permanent dipole moment associated with polar molecules (this is a ground state property). The electric transition dipole moment is given the symbol μ. When $\mu \neq 0$, the transition is said to be allowed with $\varepsilon = 100{,}000$ to 5000.

If the excitation does not involve a translation of charge, the excitation will be forbidden with $\mu = 0$ (formally) leading to $\varepsilon < 200$. Thus, an electron in the lone pair orbital (*n*) of a ketone is excited to the π^* orbital with a rotation of charge (Fig. 4). There is no resultant translation of charge, and the extinction coefficient is accordingly very small, $\varepsilon \sim 20$.

Consider a typical benzoate derivative such as that shown in Fig. 5. There is a weak transition (not allowed: cf toluene) at 263 nm with $\varepsilon \approx 200$. This is a $\pi \rightarrow \pi^*$ of the *p*-chlorophenol chromophore. At 240 nm, there is a strong, electric dipole allowed, charge transfer transition associated with a linear translation of an electron between the carboxyl group and the phenyl ring over a distance of 1 Å. The absorbance spectrum resulting from these properties is shown, in a stylized fashion, in Fig. 6.

In the vibrational infrared, the situation is similar. Thus, the simple stretching vibrations associated with the C=O and the O–H groups are fully allowed with $\varepsilon_{C=O} = 100$ and $\varepsilon_{O\text{–}H} = 10$. Infrared extinction coefficients are generally smaller than those associated with allowed electronic transitions because: (1) the charge displacements must necessarily be smaller than 0.1 Å—bond stretches of the order of 1 Å would lead to bond breaking, and (2) the dimension of the excitation is less comparable to the wavelength of the radiation (further from true resonance).

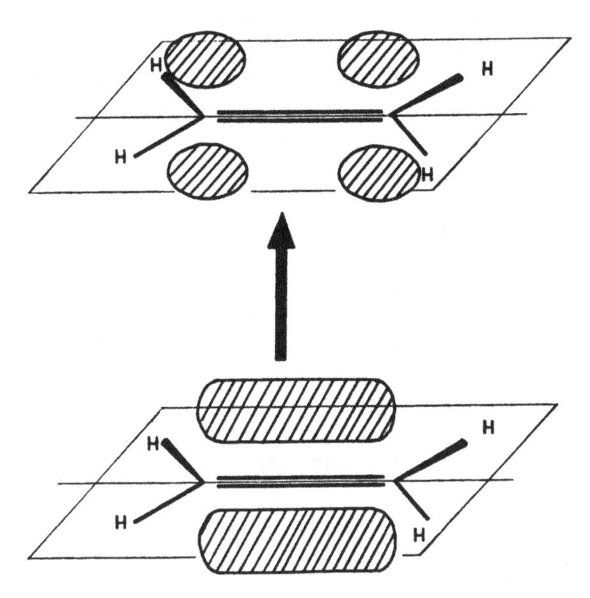

Fig. 3. Absorption of light polarized in the yz plane results in promotion of an electron from the C–C bonding orbital to the π^* antibonding orbital. In the π orbital, the charge is concentrated near the center of the bond, whereas in the antibonding orbital, there is a node between the carbon atoms, and the charge is further out. This translation of charge gives rise to an electric transition dipole, and so the transition is allowed, with $\varepsilon \approx 10{,}000$.

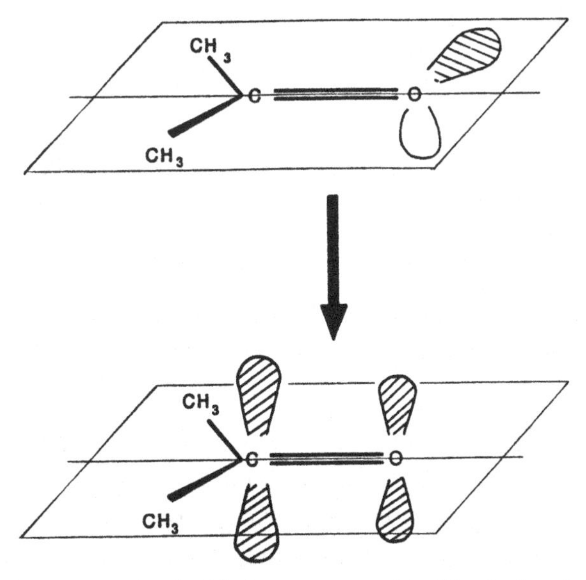

Fig. 4. The charge associated with the oxygen *n* orbital lies in the plane shown, and the transition to the $n \rightarrow \pi^*$ orbital results primarily in rotation of the charge about the C–O bond. As a consequence, the extinction coefficient is small ($\varepsilon_{290} \approx 20$).

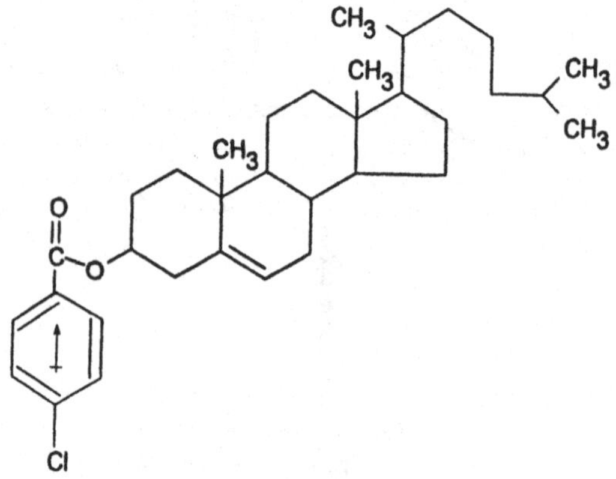

Fig. 5. A *p*-chlorobenzoate derivative of cholesterol.

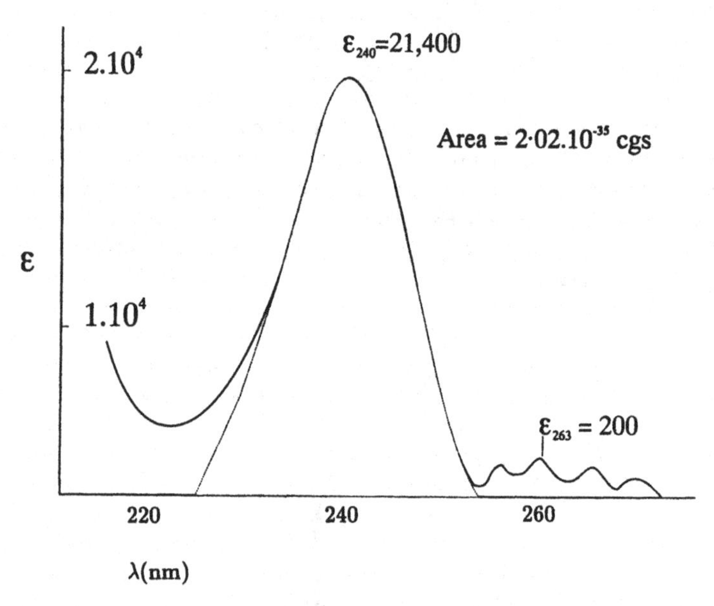

Fig. 6. The absorption spectrum of the benzoate derivative shown in Fig. 5. The area under the intense band with λ_{max} at 240 nm gives its dipole strength, *D*.

Vibrations that do not involve changes in dipole moment between the ground and excited states (transition electric dipole moment, $\mu = 0$), such as torsional modes, will accordingly be forbidden with small ε values <1.0, although such transitions are usually observed in Raman spectroscopy, which has different selection rules determined by polarizability (*see* Chapter 15, vol. 17 of this series).

3.2.2. Spin Selection Rules

An electronic transition must not involve inversion of electron spin on excitation. This was assumed above. However, associated with every ordinary electronic transition, there will be at lower energies (longer wavelengths) a corresponding excitation with a flip of the electron's spin to what is known as the triplet state of this level. This absorption is severely forbidden with very small ε values and is generally hidden under the envelope of allowed transitions. They are virtually impossible to observe, and only the triplet of the lowest energy absorption is normally detectable. Transitions back to the ground state also require an inversion of spin and are formally forbidden, and the excited state may therefore have an extended lifetime before radiative emission can occur. This is phosphorescence. Spin-forbidden transitions are most often seen in the visible spectrum of metal complexes, particularly lanthanides, with ε values <1.0.

Selection rules are guidelines. In reality, the mixing of molecular orbitals and the borrowing or the sharing of intensity can occur. Energy levels are rarely pure. Thus, formally forbidden transitions can be observed sometimes with relatively large ε values, particularly if they are close in energy to fully allowed excitations.

It is important to remember that a chromophore will have more than one electronic transition associated with it. Thus, transition metal complexes have weak d-d transitions ($\varepsilon < 100$) in the visible responsible for their color with strong allowed transitions associated with charge transfer or ligand excitations in the UV. In some cases, for example the heme groups, the electronic spectrum is totally dominated by ligand-associated excitations, which are in turn responsible for the intense color. The colors associated with nonmetallic organics (dyes, chlorophylls, carotenes, and so forth) generally derive from very strongly allowed $\pi \rightarrow \pi^*$ transitions with $\varepsilon > 20{,}000$. The hue or shade of the color is dictated by the wavelength of the excitations, and the intensity is controlled by the molar extinction coefficient.

4. Instrumentation: Absorption Spectrometers

In principle, the role of the optical absorption spectrometer is to determine the extent to which light of different wavelengths is absorbed by a sample. Today this is achieved in one of four ways:

1. Stable single-beam spectrometers;
2. Double-beam spectrometers;
3. Diode array spectrometers; and
4. Fourier transform spectrometers.

The two major features that govern the quality (and price) of an optical spectrometer are the light throughput and the spectral resolving power of the monochromator. Expensive instruments have two monochromators in series that provide a highly monochromatic (low stray light) output with narrow spectral bandwidth. In terms of performance three aspects are important:

1. High sensitivity measurements of very low absorbances (e.g., chromatography detection) generally require high light throughput at the expense of spectral resolution and stray light. Large spectral bandwidths of 1–2 nm (up to 10 nm with filters) are normally sufficient for biological samples.
2. High absorbance measurements ($A > 2.0$) are critically dependent on the absence of stray light, which ought to be <0.01%. This requires either a very high-quality diffraction grating or, better, a double monochromator. The effects of stray light are illustrated in Fig. 7. Stray light can lead to underestimation of high absorbances and the appearance of false maxima in spectra.
3. High spectral resolution also requires a double monochromator to ensure that the spectral bandwidth is less than, typically, 0.1 nm. In biological electronic spectroscopy, the absorption bands encountered are usually broad and a relatively low instrument spectral resolution (1–2 nm, wide instrument slit width) is sufficient. High spectral resolution is needed to study fine structure, such as found with the 260-nm absorption of phenylalanine in calmodulin or the sharp, atomic-like absorptions associated with the f-f transitions of lanthanide ions used as alkali metal probes.

4.1. Stable Single-Beam Spectrometers

As the name implies, a single detector monitors the intensity of a single light beam coming from the monochromator. This is illustrated in Fig. 8.

The attractive simplicity of this layout has been made practicable only relatively recently by the technological advances in spectrometer components (power supplies stabilized with feedback servo systems, more stable light sources and detectors, stable signal handling circuitry) and the ready availability of computers. In an ideal experiment, a spectrum is first measured with no cell in the measurement compartment. A

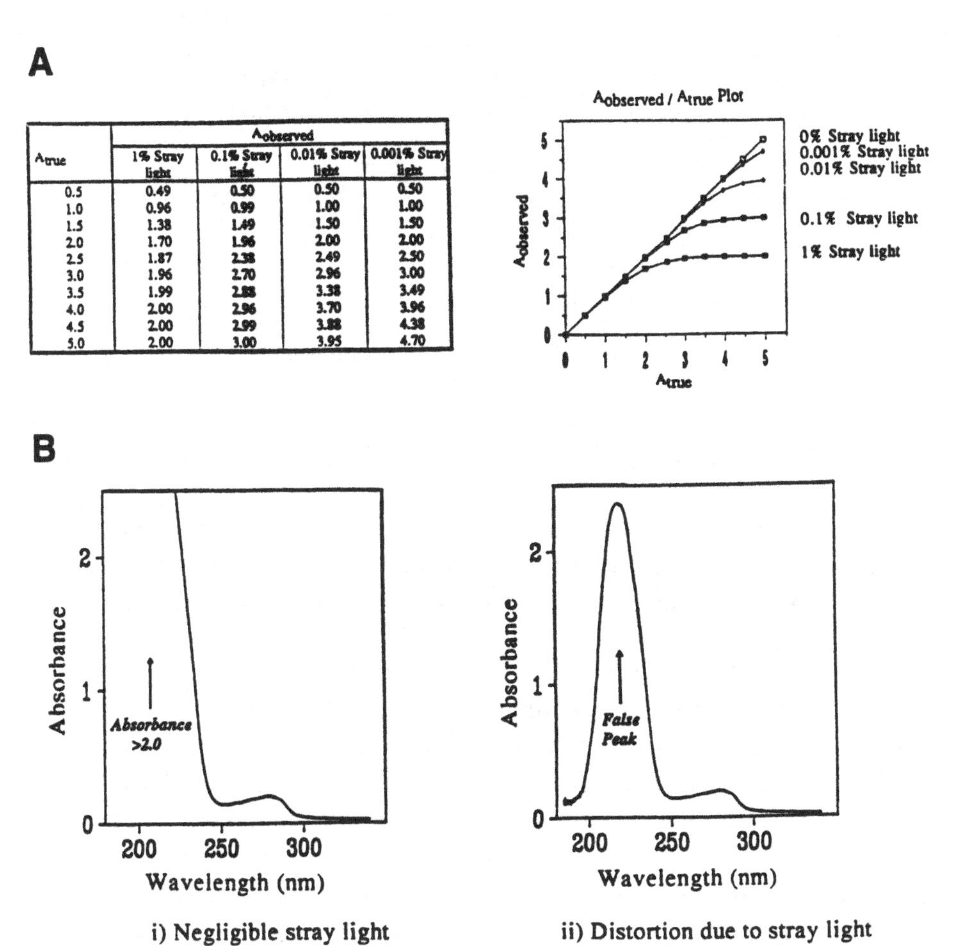

A_{true}	$A_{observed}$			
	1% Stray light	0.1% Stray light	0.01% Stray light	0.001% Stray light
0.5	0.49	0.50	0.50	0.50
1.0	0.96	0.99	1.00	1.00
1.5	1.38	1.49	1.50	1.50
2.0	1.70	1.96	2.00	2.00
2.5	1.87	2.38	2.49	2.50
3.0	1.96	2.70	2.96	3.00
3.5	1.99	2.88	3.38	3.49
4.0	2.00	2.96	3.70	3.96
4.5	2.00	2.99	3.88	4.38
5.0	2.00	3.00	3.95	4.70

Fig. 7. Stray light characteristics and the effect of stray light on the spectrum of a protein. (**A**) Table and plot of the relationship between observed and true absorbance at different levels of stray light; where absorbance is strong ($A > 2.0$), stray light must be kept low (<0.01%). (**B**) The UV absorption spectrum of human serum albumin; (i) with negligible stray light and (ii) with spectral distortion owing to stray light, giving the misleading impression that an absorbance maximum exists at about 220 nm.

reference (the solvent in the same cell as will be used for the sample) is now scanned followed by the sample (analyte + solvent). The detector registers the variation of light intensity with wavelength. The three results are stored in the memory of an on-line computer, which now holds the information illustrated in Fig. 9. Computer data manipulation turns these into the more familiar absorption spectrum. The no-cell

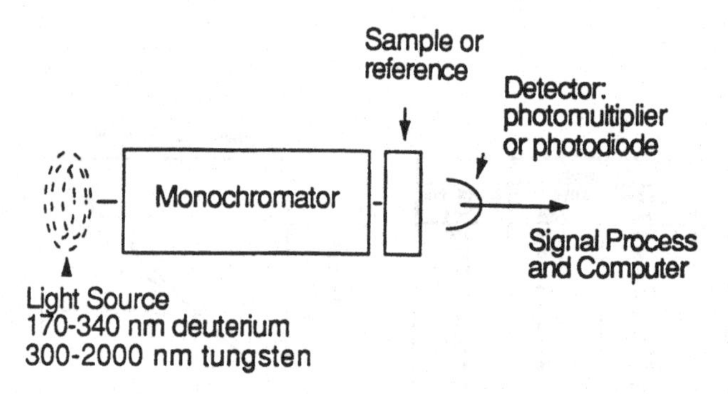

Fig. 8. The stable single-beam spectrometer.

measurement may need to be run only once a day acting as a reference for the rest of the day.

Instruments of this type require very stable light sources and detectors, and need to be relatively insensitive to background fluctuations. This is generally only feasible in the UV, Vis, and perhaps near-IR regions.

4.2. Double-Beam Spectrometers

In the double-beam spectrometer (Fig. 10), the beam coming from the monochromator is split into two; one acts as a reference beam giving a continuous measure of the intensity of the incident light (I_0) whereas the other passes through the sample for the measurement of transmitted light (I_t). A mechanical chopper carrying a mirror can be employed to send the beam alternately along two paths, or more recently, the Perkin-Elmer λ-2 spectrophotometer has been constructed with an optical beam splitter dividing the beam into two for eventual detection by two separate matched silicon photodiodes. If the reference beam is "hidden" within the spectrometer or kept effectively sealed, the instrument behaves essentially as a very stable single-beam instrument.

4.3. Diode Array Spectrometers

In the above types of spectrophotometers, each wavelength is preselected by a monochromator, and the decrease in intensity as the light beam passes through a sample is measured by a single detector. The sample can, however, be placed before the monochromator, and the spectrometer analyzes the light profile after the sample has modified it. Traditionally in UV/Vis work, this has not been popular because of the potential photochemical effects of shining "whole light" on a

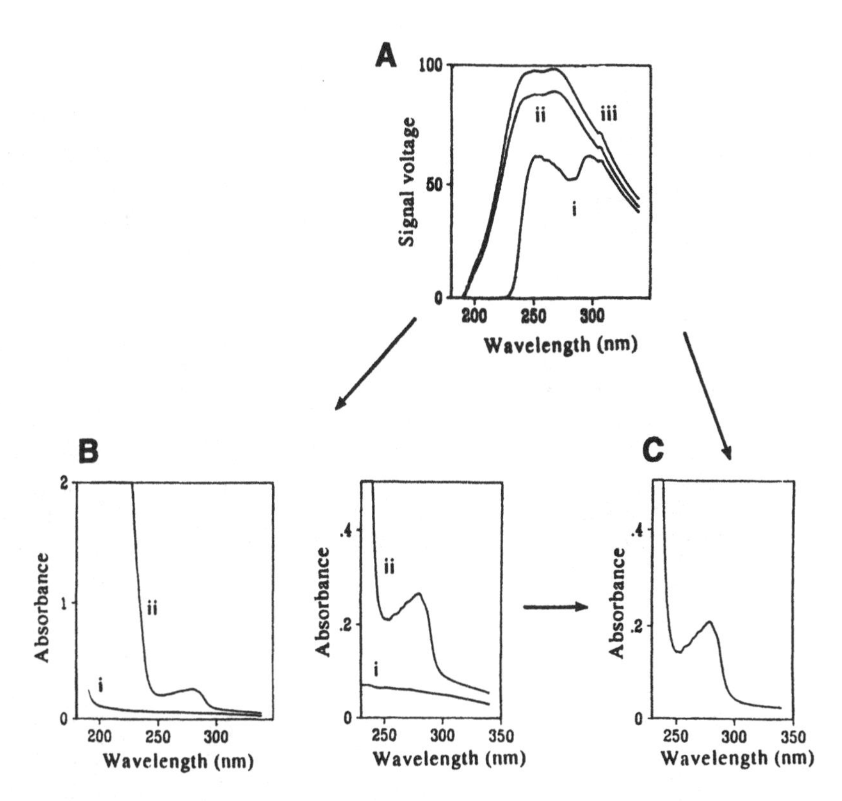

Fig. 9. Detector signal and absorption measurement. **(A)** Three sets of data are available in the computer as light intensity variation (proportional to detector signal) with wavelength; (i) the instrumental background, (ii) the cell with solvent, and (iii) the cell with analyte dissolved in the solvent. At every wavelength, the following relationship holds:

$$[\log(V_0) - \log(V_t)] = (\log[I_0] - \log[I_t]) = \log(I_0/I_t) = A$$

The proportionality constant $V = kI$ is set in the instrument construction. **(B)** The data sets can be processed in pairs as instrument background with solvent (i), and instrument background with analyte and solvent (ii), to provide two absorption vs wavelength data plots that can be compared to give the resultant absorbance spectrum of the analyte. **(C)** Processing solvent with analyte plus solvent gives the analyte spectrum alone, without the opportunity to study the quality of the solvent absorption. The remaining offset of the spectrum from zero is owing to light scattering.

sample. Judicious optical prefiltering can reduce this. In the infrared, this mode is preferred, since it avoids artifacts resulting from unabsorbed, background, black-body radiation from the monochromator itself. A new design of UV/Vis spectrometer offered by Pharmacia (Uppsala, Sweden) places the sample before the monochromator with the justi-

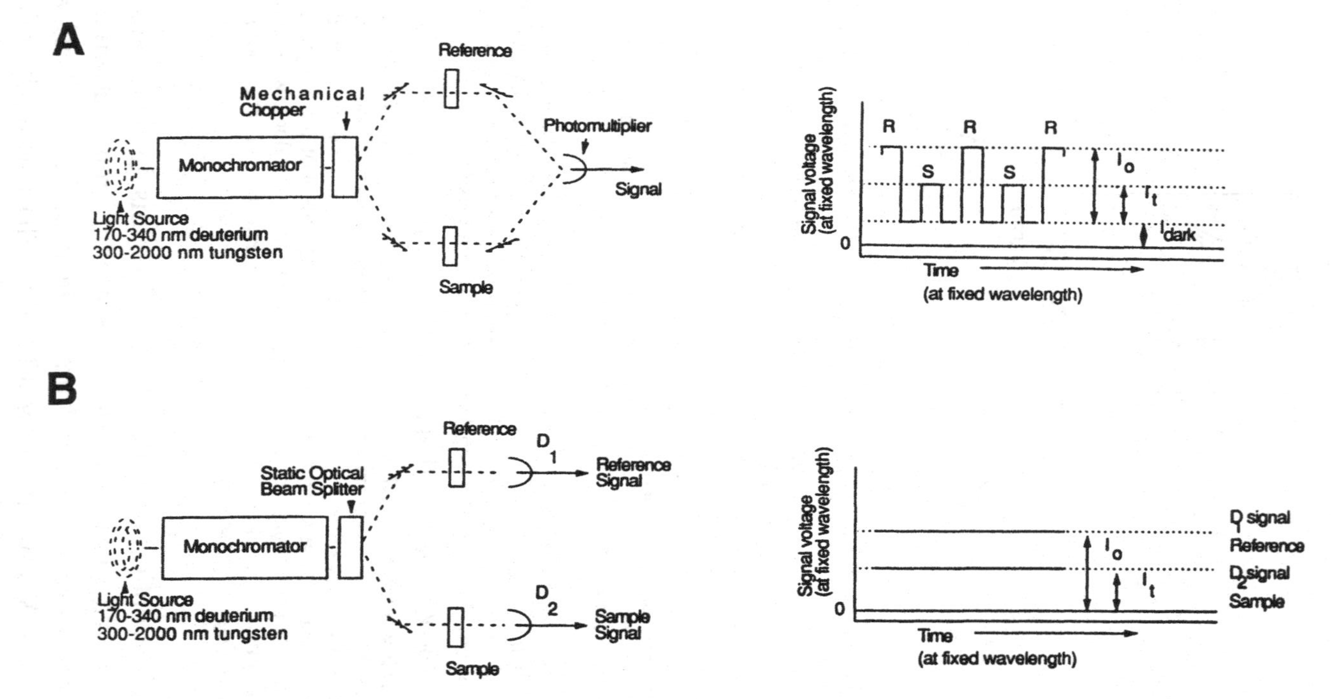

Fig. 10. Double-beam spectrometers. **(A)** Production of the double beam using a mechanical chopper with one detector; the chopper diverts a monochromatic light beam alternately through the reference and sample compartments to a single detector (photomultiplier). R = light passing through the reference compartment period; L = light passing through the sample compartment period; D = dark period, in which the signal level is normally negligible in the UV and visible regions, but must be taken into account for IR work. **(B)** A static optical beam splitter with two detectors. D_1 and D_2 are matched Si photodiodes.

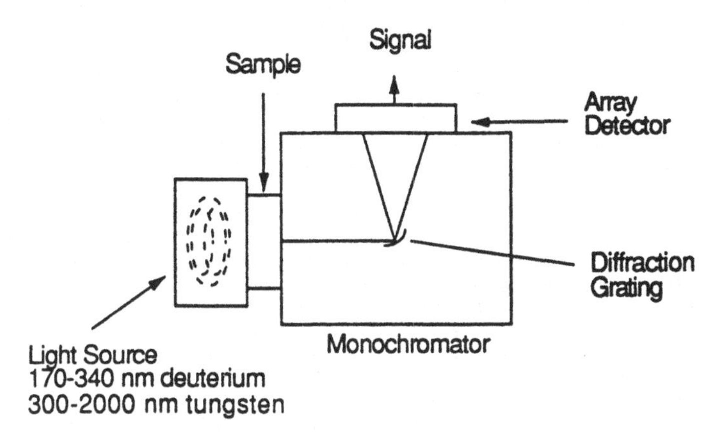

Fig. 11. The diode array spectrometer.

fiable advantage that stray light is less critical even when the instrument is operated with an "open" cell compartment.

In diode array spectrometers (Fig. 11), the profile of whole light is analyzed before and after transmission through the sample and subsequent dispersion by a monochromator. The array detector is effectively an electronic photographic plate with photodiode elements individually responsible for monitoring wavelength regions on the order of 1 nm (cf the grain of a photographic plate), which means that wavelength resolution is generally lower than that associated with scanning instruments.

This design offers many advantages:

1. Every wavelength is monitored simultaneously, permitting much faster spectrum acquisition compared with scanning systems (seconds rather than minutes). Whole spectra can be measured in seconds, on-the-fly during chromatography or to follow the course of a reaction.
2. Accumulation time can be increased to make the diode array detector apparently more sensitive than the scanning system for a given measurement time.
3. Perhaps the most important advantage of diode array is that every wavelength provides data equally "good" for analysis. In scanning measurements, the uncertainty of determining absorbances on the side of absorption bands restricts analysis to measurements made at peaks, troughs, or shoulders. Diode array data allow the use of linear regression methods to be applied more faithfully to very many points in a spectrum (e.g., every 1 nm in the range 340–240 nm [101 points]).

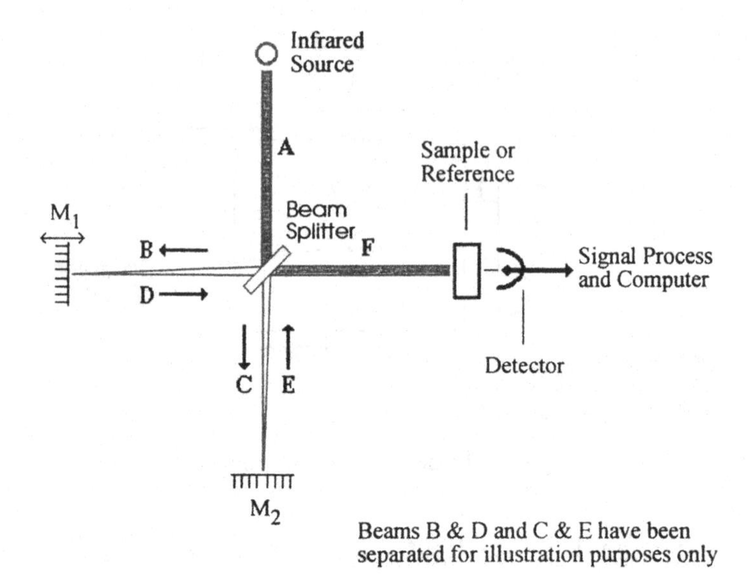

Fig. 12. The Fourier transform (FT) spectrometer. In an interferometer, the collimated polychromatic beam A from an IR source is split into two beams, B and C. These beams are reflected back by mirrors M_1 and M_2 as beams D and E, respectively, which reconstruct to form a single beam, F, that is incident on the sample. If the distances from the beam splitter to the mirrors M_1 and M_2 are identical, the reconstructed beam will have the characteristics of beam A. If the mirror M_1 is set to oscillate along the optic axis as illustrated, the distance traversed by beams B and D will be that traversed by beams C and E. The time to travel these various distances will be dependent on the wavelength of light. The different wavelength components of the beams D and E will now arrive back at the beam splitter for reconstruction at different times. The same wavelength components of these two beams will interfere with one another either constructively or destructively, so that the intensity of beam F will vary with time as a function of the different path lengths B + D and C + E, which are controlled by the mirror M_1. Computer mathematical analysis, based on Fourier transform techniques, of the time dependence of the beam F intensity will allow the extraction of data relating to intensity vs wavelength. Measurements with and without samples provides the absorption spectrum as a single-beam measurement.

These instruments are expensive, generally have low spectral resolution, and employ photodiode detection that is noisier than the photomultiplier for low-light-level applications. The same requirements for power supply and lamp stability apply as for single-beam scanning instruments. Nevertheless, given lower cost, they would probably be more readily employed.

4.4. Fourier Transform Spectroscopy

Fourier transform techniques have many advantages over the traditional monochromator-based methods: they detect all wavelengths simultaneously, giving better signal accumulation per unit time and, without monochromators, have reduced light losses. However, these advantages become less apparent when the signal is light-level limited (not detector or background radiation limited, as in the IR). In the UV/Vis the oscillating mirror must be moved back and forth with an accuracy of angstroms rather than microns that leads to the need for ultraprecise motion and great demands on computing. These instruments are therefore generally prohibitively expensive and only usually viable for the high spectral resolution demands (the major benefit of FT technique) of atomic absorption and fluorescence spectroscopy. They have not yet been considered for the relatively undemanding requirements of biological UV/Vis spectroscopy. The design of FT spectrometers for use in the IR is outlined in Fig. 12; their application is described in Chapter 14 of this volume.

CHAPTER 13

The Measurement of Electronic Absorption Spectra in the Ultraviolet and Visible

Alex F. Drake

1. General Principles

There is a general trend, for both instrumental and computational reasons, towards spectra recorded on single-beam or effectively single-beam spectrometers. The results of all measurements ought to be treated as "spectra" with the analyte absorption spectrum being the difference between the (analyte + solvent) spectrum and the solvent spectrum. Double beam spectrometers lend themselves to the traditional methodology involving matched cells, one containing solvent only in the reference beam and the other containing (analyte + solvent) in the sample beam. This was attractive, since continuous reference to solvent provided chart-recorder presentations that were easier to process by hand because the solvent absorption was corrected for automatically. In regions where the solvent is absorbing strongly, reference correction may be carried out with absorbance values from the side of a solvent absorption band; inevitably, the absorption/wavelength data here are imprecise, particularly when manual data manipulation is used. On the other hand, continuous compensation with matched cells leads to the loss of the solvent-in-cell spectrum, which can be an important feature of absorption spectrum measurement validation.

The following scheme is recommended for any spectrum measurement:

From: *Methods in Molecular Biology, Vol. 22: Microscopy, Optical Spectroscopy, and Macroscopic Techniques* Edited by: C. Jones, B. Mulloy, and A. H. Thomas

1. Set air zero. With only air (or nitrogen) and any beam stops or special cell holders in the measurement beam, the lowest absorbance reading in the wavelength scan of interest should be offset to zero absorbance. This is normally at the starting wavelength for a from-long-to-short wavelength scan.
2. Run spectrum of cell containing solvent (store result in computer memory).
   a. At the beginning of a spectrum (e.g., 340 nm for a protein solution), where neither solvent nor analyte absorbs, an apparent absorbance of about 0.05 (due to reflection) ought to be observed for a good-quality clean cell containing "clean" solvent.
   b. Based on prior knowledge or literature, the spectrum of the solvent should be truly characteristic of the solvent (e.g., 1 cm of distilled water should not have appreciable absorption at wavelengths >200 nm; the far UV absorption spectrum is an excellent measure of the purity of distilled water).
   c. The spectrum of the solvent will define the transmission range for subsequent measurements.
3. Run spectrum of cell containing (analyte + solvent), and store in computer memory.
4. Reject data points in steps 2 and 3 that correspond to absorbance values greater than those permitted by the instrument (generally solvent data with $A > 1.0$ and [analyte + solvent] data with $A > 2.0$). In CD spectroscopy, absorbance values greater than about 1.5 are not recommended; for wavelengths below 210 nm, absorbances in the range 0.5–1.0 are required. The ideal value for both ordinary UV/Vis absorption and CD spectroscopy is $A = 0.864$. In the infrared, the ideal value of the absorbance is lower at about 0.4.
5. Print the analyte absorption spectrum after data manipulation (solvent correction, light scattering correction, concentration/pathlength correction for extinction coefficients, and so on [Fig. 1]). Although presented for computer data manipulation, this methodology is also recommended for traditional chart-recorder outputs (*see* Fig. 1 and Chapter 12, Fig. 2). If background solvent absorption is rising steeply and awkward to correct for by hand, then the zone of artifact is being approached, and the data are already becoming unreliable and ought to be rejected.

   Other than the form of the spectrum itself there are two other important aspects of a spectrum.

   a. Outside the absorption wavelength range of the corrected analyte spectrum, prior to the leading edge, the absorbance should be zero. If not, the sample is either scattering light, there is impurity present, or the cell has not been properly cleaned, filled, or positioned. Light scattering can

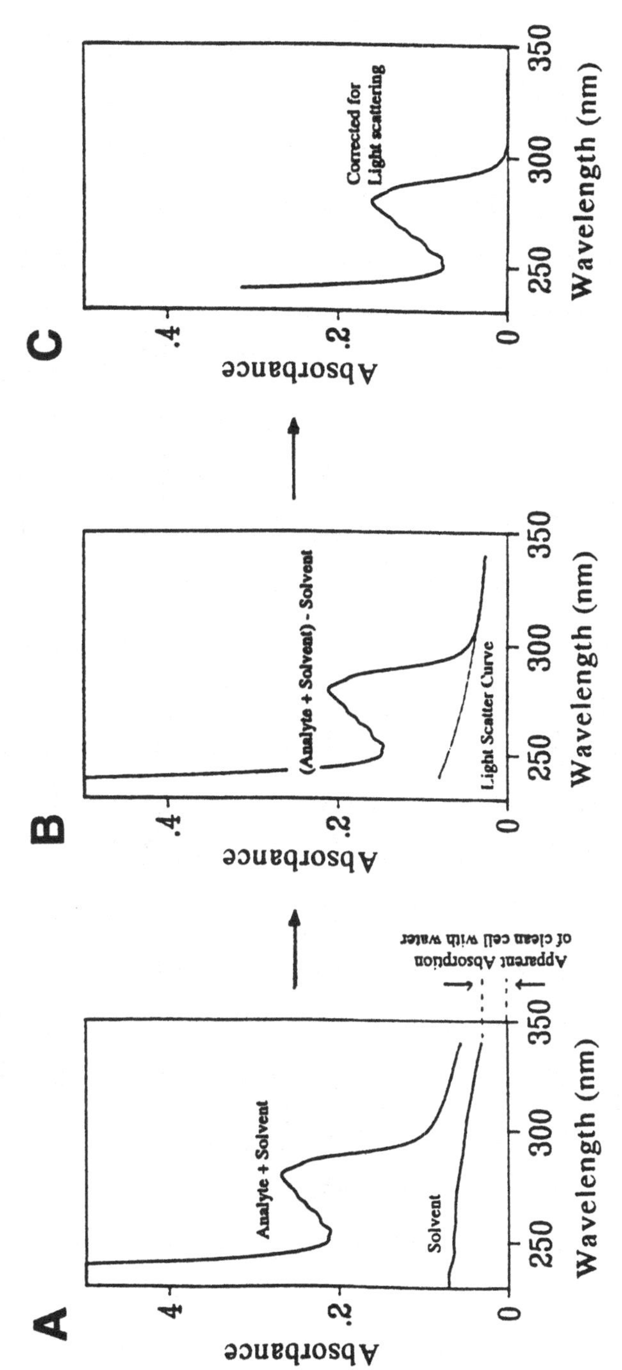

Fig. 1. Stages in the correction of raw absorbance data to the final absorbance spectrum of the analyte, for an aqueous solution of serum albumin. (A) Spectra as recorded of the solvent and of the sample (analyte + solvent). (B) Correction of the sample spectrum by subtraction of the solvent spectrum. The offset of this spectrum from zero is owing to light scattering. (C) Correction of the spectrum for light scattering: *See* Section 2.3.

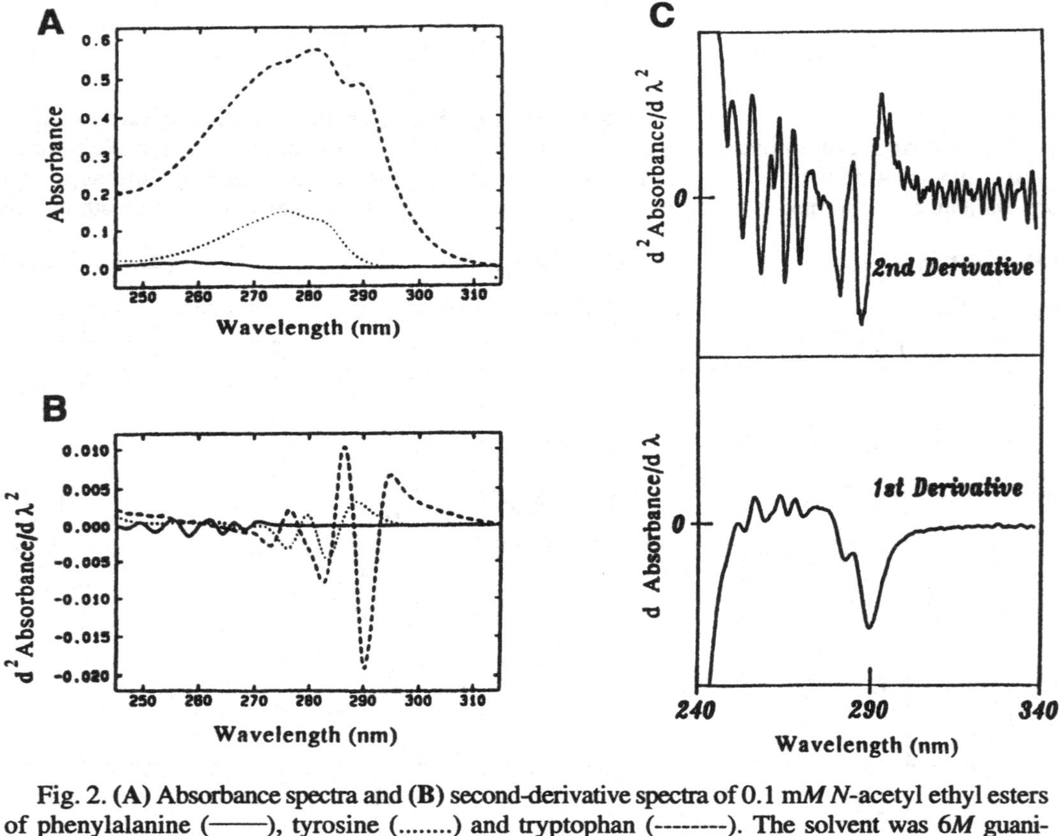

Fig. 2. **(A)** Absorbance spectra and **(B)** second-derivative spectra of 0.1 m*M N*-acetyl ethyl esters of phenylalanine (——), tyrosine (........) and tryptophan (--------). The solvent was 6*M* guanidine—20 m*M* potassium phosphate, pH 6.5. Data reproduced, with permission, from ref. *9*, copyright (1982) American Chemical Society. **(C)** Derivative spectra of human serum albumin (from the absorbance data in Fig. 1C).

be compensated for mathematically, or the sample should be centrifuged or filtered.

b. A spectrum can be analyzed not only in terms of the wavelength and intensity of spectral features (maxima, minima and shoulders) but also in terms of the relative intensities of spectral features. The latter is a particularly useful diagnostic. The ratio of two absorbances at different wavelengths provides a dimensionless constant, independent of concentration and cell pathlength, characteristic of the analyte:

$$(A_{\lambda 1}/A_{\lambda 2}) = (\varepsilon_{\lambda 1}/\varepsilon_{\lambda 2}). \quad (1)$$

The ratio of the absorbances of a nucleic acid at the maximum (typically A_{260}) and minimum (typically A_{240}) can act as a good reference of identity and purity. Nucleic acid contamination of proteins can be judged from the A_{280}/A_{260} ratio. The A_{280}/A_{250} ratio of a good, pure tryptophan or tyrosine-containing protein ought to be of the order of 0.5 (cf serum albumin: *see* Fig. 1).

2. The Spectrophotometric Determination of Protein and Peptide Concentration

The determination of protein (peptide) absolute concentration is an important task, particularly for quantitative studies of association, binding, and the secondary structure analysis by CD spectroscopy. The excellent review by Darbre *(1)* is recommended, although the conclusions are based largely on this author's own experience. Three general methods are available.

2.1. Other Methods

2.1.1. Direct Dry Weight

The dissolution of a preweighed sample in a known volume of a solvent is the simplest means of producing a solution of known concentration. However, this requires a precise knowledge of the chemical integrity of the test sample. It is often difficult to account for impurities (coprecipitants and residues from a previous isolation step), for solvent of crystallization, for bound lipids, for bound carbohydrates (in glycoproteins), and so on. In controlled environments with a good six-figure microbalance, 0.2 mg is perhaps the smallest amount that can be weighed with reproducibility. Smaller quantities will require analysis of a prepared solution. Gravimetric analysis after precipitation or solvent evaporation will not give accurate results for sample quantities less than several milligrams.

2.1.2. Chemical Analysis

This inevitably leads to destructive techniques. Direct determination of nitrogen content is possible, but is both time-consuming and requires several milligrams of sample, so is therefore rarely employed nowadays. A protein can be hydrolyzed completely to its constituent amino acids with 6*M* HCl for 18–24 h at 110°C, allowing the residual amino acids to be quantified by chromatography. Tryptophan itself is destroyed in this process. However, if the protein sequence (or amino acid composition) is known, then the analysis of individual amino acids will provide several measures of the original protein concentration and is potentially the most secure of the methods described here.

Proteins (containing specific constituent amino acids) can be chemically derivatized or complexed with coloring agents, such as copper salts, to give fluorophores or UV/Vis chromophores for subsequent analysis. Typical of the latter are the methods of Lowry et al. *(2)* and Bradford *(3)*, which involve the colorimetric analysis of complexes formed with copper ions or Coomassie Brilliant Blue, respectively. These methods tend to be protein-specific, requiring previous calibration, rather than general. Even so, accuracy may not be high. The use of direct UV spectrophotometry avoids unnecessary sample handling, and saves time and trouble, especially if other spectroscopic studies (for example, by circular dichroism: *see* Chapter 16) are planned.

2.2. Direct Spectrophotometry

Proteins normally carry absorbing groups that in themselves can act as concentration monitors through spectrophotometry. In extreme cases, these include prosthetic groups, such as the heme of hemoglobin, which once calibrated provides an internal concentration index. However, more generally, the side chains of tryptophan, tyrosine, phenylalanine, and the disulfide bond act as the UV active agent.

Beer's law is written as:

$$\varepsilon = \frac{A}{[c] \cdot l}$$

or

$$\varepsilon = \frac{\mathrm{MW} \cdot A}{[x] \cdot l} \qquad (2)$$

Protein concentration is given by:

$$[c] = \frac{A}{\varepsilon \cdot l}$$

or

$$[x] = \frac{\mathrm{MW} \cdot A}{\varepsilon \cdot l} \tag{3}$$

where $[c]$ is the protein concentration in mol/L, $[x]$ is the concentration in mg/mL, MW is the mol wt and l is cell pathlength.

For absorbance measurements at 280 nm, the contribution from phenylalanine is negligible, and if light scattering is corrected for, the observed value may be considered to be an algebraic sum of contributions from individual tryptophans, tyrosines, and disulfides:

$$\begin{aligned} A_{280} &= n_{trp} A_{280}^{trp} + n_{tyr} A_{280}^{tyr} + n_{s-s} A_{280}^{s-s} \\ &= [\varepsilon_{trp}\, n_{280}^{trp} + \varepsilon_{tyr}\, n_{280}^{tyr} + \varepsilon_{s-s}\, n_{280}^{s-s}] \cdot [c] \cdot 1 \end{aligned} \tag{4}$$

Rearrangement gives the concentration formula for a 1-cm cell:

$$[c] = \frac{A_{280}}{\varepsilon_{280}^{trp} \cdot N_{trp} + \varepsilon_{280}^{tyr} \cdot n_{tyr} + \varepsilon_{280}^{s-s} \cdot n_{s-s}} \tag{5}$$

where $\varepsilon^{x}{}_{280}$ and n_x are the extinction coefficients and the number of respective residues in the protein. In CD spectroscopy, data are reported per amino acid residue with the concentration expressed as the protein mean residue concentration in residue mol/L; $[C]_r = [C]/n$ where n is the total number of residues. $[C]_r$ is equal to $\mathrm{MW}_r/[x]$, i.e., the mean residue mol wt of the protein (typical value 113) divided by the total protein concentration in mg/mL. In this way, data are presented independent of protein length (the total number of amino acids in the protein).

This general method was advocated by Wetlaufer *(4)* in 1962. However, it does require knowledge of the values of ε_{280} for tryptophan, tyrosine and the S–S disulfide. The precise values of these extinction coefficients are dependent on the environment of the chromophores, i.e., whether they are exposed to the polar solvent or buried in the relatively hydrophobic nonpolar core of the protein. Two recent papers have discussed this at length in an attempt to obtain consensus values.

To overcome the uncertainty of environment, Gill and Von Hippel *(5)*, following the lead of Edelhoch *(6)*, advocate that the protein (or protein solution) is first quenched into 6.0*M* guanidinium hydrochloride, and 0.02*M* phosphate buffer, pH 6.5. The protein is then considered to be denatured with all aromatics exposed, and only now is the observed A_{280} an algebraic sum of the three components. A correlation of the ε_{280} values of 18 "normal" globular proteins and comparison with data for *N*-acetyl-L-tryptophanamide, Gly-L-Tyr-Gly, and Cystine all in the denaturing condition provide, after correction for light scattering, the 1-cm cell formula:

$$[c] = \frac{A_{280}}{5690 \cdot n_{trp} + 1280 \cdot n_{tyr} + 120 \cdot n_{s-s}} \quad (6)$$

Mach et al. *(7)* have applied a "matrix linear regression procedure and a mapping of average absolute deviations between experimental and calculated values to find extinction coefficients (ε_M, 1 cm, 280 nm) of $5540M^{-1}$ cm^{-1} for tryptophan and $1480M^{-1}$ cm^{-1} for tyrosine residues in an 'average' protein, as defined by a set of experimentally determined extinction coefficients for more than 30 proteins." This latter method requires the protein to be dissolved simply in 50 m*M* sodium phosphate buffer containing 0.02% sodium azide (pH 7.0). The concentration formula for a 1-cm cell now reads:

$$[c] = \frac{A_{280}}{5540 \cdot n_{trp} + 1480 \cdot n_{tyr} + 134 \cdot n_{s-s}} \quad (7)$$

Mach et al. *(7)* state that their values are the most representative since Wetlaufer's attempt to correlate UV absorption with the concentration of proteins gives a 10% error, whereas the Gill and Von Hippel values *(5)* of the extinction coefficients lead to a 5% standard deviation between experimental and calculated ε_{280} values for the studied proteins. A survey of the tables in refs. 5 and 7 presenting standard deviation values shows that, although the Mach et al. values *(7)* are overall better than those of Gill and Von Hippel *(5)*, there are examples where the converse is true. According to either Eq. (6) or (7), insulin presents a large discrepancy between calculation and experiment. The denaturing conditions of Gill and Von Hippel *(5)* give the better results, which underline the conclusion that a special situation applies in this case requiring nonstandard values. Insulin does not contain tryptophan, and it is noteworthy that its UV maximum is 2 nm blue shifted compared to the standards, reinforcing the view that the tyrosines are located in special environments.

2.3. Light Scattering Correction for Proteins

As the size of a molecular assembly approaches the wavelength of incident light, so the light scattering becomes greater and the observed A_{280} will be an overestimate. The effect is proportional to λ^{-4}. From a practical point of view, light scattering is best assessed at long wavelengths prior to the onset of true electronic absorption. A plot of the expression:

$$\log(A_{s,\lambda}) = \text{slope} \cdot \log(\lambda) + \text{intercept} \qquad (8)$$

over the wavelength range 350 nm to 320 nm will allow extrapolation of the values $A_{s,\lambda}$, the apparent absorbance owing to light scattering, to the wavelengths of interest under the true electronic absorbance. Specifically at 280 nm, $A_{s,280}$ is given from an analysis of Eq. (8) as:

$$A_{s,280} = 10^{(2.5\log A_{320} - 1.5\log A_{350})} \qquad (9)$$

Thus

$$A_{280} \Rightarrow A_{280} - A_{s,280} = A_{280} - 10^{(2.5\log A_{320} - 1.5\log A_{350})} \qquad (10)$$

This expression can be applied generally; for example, the numerators of Eqs. (6) and (7) can be modified accordingly. The latter provides the total protein concentration in a 1-cm pathlength cell:

$$[c] = \frac{A_{280} - 10^{(2.5\log A_{320} - 1.5\log A_{350})}}{5540 \cdot n_{trp} + 1480 \cdot n_{tyr} + 134 \cdot n_{s-s}} \qquad (11)$$

The situation is further complicated with oligopeptides dissolved in nonaqueous solvents, such as methanol or trifluoroethanol. It is well known that on changing solvent from water, UV absorption changes will be observed particularly associated with tyrosine residues (Solli and Herskovits *[8]*). These changes, which may be related to both the change in peptide conformation and the solvent environment, have not as yet been critically reviewed.

3. Derivative Spectroscopy and Multicomponent Analysis

Presenting spectra (Fig. 2) as the first derivative, $dA/d\lambda$ vs λ or as the second derivative $d^2A/d\lambda^2$ vs λ offers an alternative approach to data analysis. Beer's law can be applied equally as well to the peaks and troughs of derivatives curves as it can to the features of the ordinary absorbance spectrum, although the extinction coefficient will be different from that found in the ordinary spectrum. This has been discussed by Levine and Federici *(9)*, who forward two advantages: (1) Derivative spectroscopy improves the resolution of the absorption spectrum to the extent that contributions from tryptophan, tyrosine, and phenylalanine can be individually assessed. (2) The second derivative spectrum in particular is insensitive to the slowly changing absorbance profiles in the 320–240 nm region associated with disulfide absorption and light scattering. Analysis can therefore proceed neglecting these contributions.

Multicomponent analysis based on the application of Beer's Law at several wavelengths ought to improve precision. Levine and Federici *(9)*, employing a diode array spectrophotometer, have applied this concept to the analysis of second-derivative spectra derived from an ordinary spectrum composed of some 50 points in the 320–240 nm region. Although the results were encouraging, enabling the differentiation of the three aromatic amino acids, the method was seen to be very sensitive to the precise nature of the reference spectra.

These methods have not become general and perhaps deserve greater attention particularly in view of the wider availability of computers.

References

1. Darbre, A. (1986) Analytical methods, in *Practical Protein Chemistry—A Handbook* (Darbre, A., ed.), Wiley, New York.
2. Lowry, O. H., Rosebrough, N. J., Fair, A. L., and Randall, R. J. (1951) Protein measurement with the Folin phenol reagent. *J. Biol. Chem.* **193,** 265–275.
3. Bradford, M. M. (1976) A rapid and sensitive method for the quantitation of microgram quantities of protein utilizing the principle of protein-dye binding. *Anal. Biochem.* **72,** 248–254.
4. Wetlaufer, D. B. (1962) Ultraviolet spectroscopy of proteins and amino acids. *Adv. Protein Chem.* **17,** 375–378.
5. Gill, S. C. and Von Hippel, P. H. (1989) Calculation of protein extinction coefficients from amino acid sequence data. *Anal. Biochem.* **182,** 319–332.
6. Edelhoch, H. (1967) Spectroscopic determination of tryptophan and tyrosine in proteins. *Biochemistry* **6,** 1948–1954.
7. Mach, H., Middaugh, C. R., and Lewis, R. V. (1992) Statistical determination of the average values of the extinction coefficients of tryptophan and tyrosine in native proteins. *Anal. Biochem.* **200,** 74–80.
8. Solli, N. J. and Herskovits, T. T. (1973) Solvent perturbation studies and analysis of protein and model compound data in denaturing organic solvents. *Anal. Biochem.* **54,** 370–378.
9. Levine, R. L. and Federici, M. M. (1982) Quantitation of aromatic residues in proteins: Model compounds for second-derivative spectroscopy. *Biochemistry* **21,** 2600–2606.

CHAPTER 14

Analysis of Polypeptide and Protein Structures Using Fourier Transform Infrared Spectroscopy

Parvez I. Haris and Dennis Chapman

1. Introduction

Infrared spectroscopy was one of the earliest techniques to be used for the structural studies of polypeptides and proteins *(1,2)*. However, a major difficulty that limited earlier studies of such biological molecules was the absorption of liquid H_2O, which shows strong absorption over much of the fundamental region of the infrared spectrum. This severely limited the analysis of biological molecules in their native state, and necessitated the use of dry films, KBr disks, or D_2O as a solvent. There is no strong absorption from D_2O, in the region 1700–1500 cm^{-1}, and this spectral region is one which is particularly important for the study of polypeptides and proteins. The infrared spectra of liquid H_2O and for comparison liquid D_2O are shown in Fig. 1. The recent development of computerized FT-IR (Fourier transform infrared) instrumentation now permits the subtraction of background water absorptions from dilute samples *(3,4)* and, hence, enables the study of biomolecules in their more natural environment, e.g., H_2O, buffer solutions, and so on. This approach has revolutionized the application of infrared spectroscopy for the study of biological molecules.

X-ray diffraction, nuclear magnetic resonance (NMR), and circular dichroism (CD) (Chapter 16, in this volume) spectroscopy are popular techniques used to obtain information about the structure of proteins

From: *Methods in Molecular Biology, Vol. 22: Microscopy, Optical Spectroscopy, and Macroscopic Techniques* Edited by: C. Jones, B. Mulloy, and A. H. Thomas

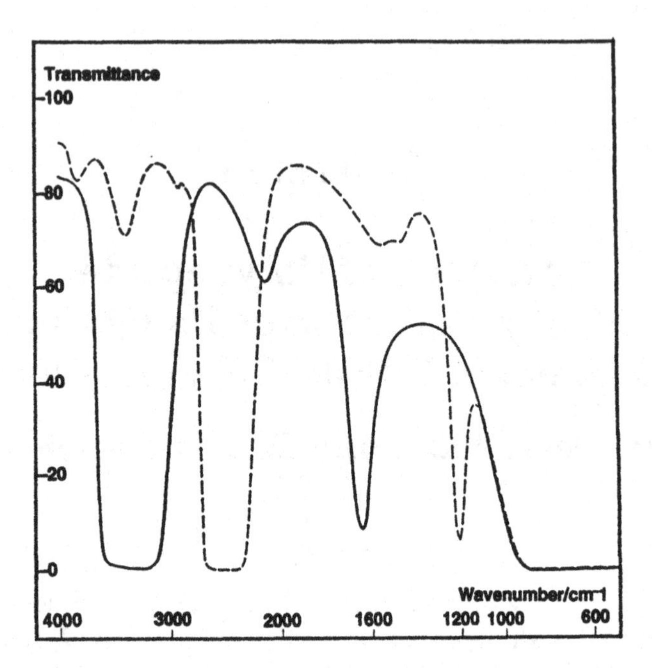

Fig. 1. FT-IR transmittance spectrum of H_2O (continuous line) and of D_2O (broken line) recorded in a calcium fluoride cell fitted with a 6-µm tin spacer. The spectra were recorded at 20°C. The peaks shift to lower wave numbers in D_2O.

and polypeptides. X-ray diffraction provides the complete three-dimensional (3D) structure of a protein in the crystalline state. However, each of these popular techniques has its own particular problems. Thus, X-ray diffraction requires large (200 µm) crystals, which are sometimes difficult to obtain. NMR spectroscopy can now be applied to obtain the complete 3D structure of a protein in solution; however, this is currently limited to the investigation of proteins of mol wt <20,000 and is severely restricted in applications to membrane proteins. CD spectroscopy can provide information about the secondary structure content of a protein, but it also has its own limitations, such as light scattering difficulties, which occur with membrane proteins. Infrared spectroscopy does not suffer from problems related to light scattering, molecular size, or crystallization. Indeed, it is possible to obtain infrared spectra of samples in solution, solid, crystalline and the gel states. In recent years, infrared spectroscopy has become a popular technique for the study of the structure of biological macromolecules, such as lipids, proteins, and polypeptides *(5,6)*.

Table 1
Characteristic Infrared Bands of the Peptide Linkage[a]

Designation	Approximate frequency, cm^{-1}	Origin
A	3300	N–H (s)[b]
B	3100	N–H (s)[b]
I	1690–1600	C=O (s)[b] 80%, N–H (b)[c] 10%, C–N (s)[b] 10%
II	1575–1480	N–H (b)[c] 60%, C–N (s)[b] 40%
III	1301–1229	C–N (s)[b] 30%, N–H (b)[c] 30%, C=O (s)[b] 10%, O=C–N (b)[c] 10%
IV	767–625	O=C–N (b)[c] 40%, other 60%
V	800–640	N–H (b)[c]
VI	606–537	C=O (b)[c]
VII	200	C–N (t)[d]

[a]The values were derived from those given by Susi *(2)*. The data were originally obtained from the infrared spectra of model compounds (e.g., *see* ref. *8*). Amides IV to VII are out-of-plane modes, i.e., out of the plane of the *trans*-CONH- grouping; the others are in-plane modes.
[b]s: stretching.
[c]b: bending.
[d]t: torsion.

1.1. The Amide Bands and the Conformation of Polypeptides and Proteins

The amide bond, present in polypeptides and proteins, gives rise to a series of infrared active vibrations known as the amide vibrations *(7,8)*. In all, there are nine such vibrations (*see* Table 1). Of these, the amide I and amide II absorption bands are by far the most useful for structural studies. Use has, however, also been made of other vibrational modes, particularly the amide III *(9)* and amide A *(10)* bands.

The amide I band arises mainly from the C=O stretching vibrations of the amide groups weakly coupled to the in-plane N–H bending and C–N stretching modes *(2,7)*. The precise frequency of the amide I band depends on the nature of the hydrogen bonding scheme involving the C=O and N–H groups. As the nature of this bonding changes with secondary structure, the frequency of the amide I band also changes. In this way, the frequency of the amide I band can be used to differentiate among different secondary structural classes, which occur in proteins and polypeptides. The relationship between the frequency of the amide I band and the type of secondary structure has been demonstrated by studies conducted with homopolypeptides that adopt well-defined secondary structures *(2,11,12)*.This has been supported by theoretical calculations of the normal modes of vibrations of the peptide groups *(13)*. Experimental studies with proteins and polypeptides of known 3D structure have further shown very good correlation between the frequency of the amide I band and the different types of secondary structure present *(14,15)*. In principle, a globular protein containing several types of substructure will give several amide I maxima, but the large half-widths of these components can prevent their resolution. Recent application of techniques, such as deconvolution and derivative procedures, have made it possible to separate and identify the highly overlapping components of the amide I band *(5,6,15,16)*. Mathematical derivative and deconvolution procedures work by artificially narrowing the existing infrared absorption bands, thereby resolving any overlapping components *(17,18)*.

Another powerful approach for the detection of small changes in protein structure is the technique of difference spectroscopy *(6)*. This involves the digital subtraction of a protein infrared spectrum in state A from that of the protein in state B. The resultant difference spectrum only reveals peaks from those functional groups within the protein that undergo changes in their structure or environment owing to the transition from state A to state B. This method has been shown to be very useful in the study of a number of proteins that undergo trigger-induced changes from one state to another (*see* ref. *6* for a review). Polarized infrared spectroscopy is a useful tool for the determination of the orientation of the different secondary structures within the lipid bilayer (*6*).

The amide II band principally arises from N–H bending vibration of the amide bond. This band is most useful for studying hydrogen-deuterium exchange of the amide groups in proteins and polypeptides (*see* Section 3.1.).

2. Instrumentation

Fourier transform infrared (FT-IR) spectroscopy relies on the principles of interferometry and Fourier transformation for its speed and sensitivity *(19)*. Traditional dispersive infrared spectrometers irradiate the sample sequentially with a narrow range of wavelengths produced by monochromation of a polychromatic beam until the entire spectral range has been scanned. FT-IR spectroscopy, on the other hand, irradiates the sample with all wavelengths simultaneously. This is achieved by directing a polychromatic beam onto a beam splitter, which directs the beam onto two mirrors at right angles to each other. One of these mirrors moves in a direction perpendicular to its axis, such that when the beams recombine at the beam splitter, a path length difference is introduced, which results in interference. The recombined beam is directed through the sample and onto the detector. The detector output as a function of mirror retardation produces an interferogram, which is the sum of the sine waves for all frequencies present. Fourier transformation of the interferogram produces the spectrum. Since monochromation is not required, the infrared beam is not attenuated, and the signal-to-noise ratio of the resultant spectrum is therefore high. The length of time required to produce the spectrum is determined by the total distance the moving mirror must travel to generate the interferogram, and not the time required to select each frequency to be directed to the sample, as is the case for dispersive instruments. This results in a significantly reduced data collection time (for spectra of equal range and resolution) for FT-IR as opposed to dispersive spectrometers. An FT-IR spectrometer consists of two parts: an optical bench containing the interferometer and a computer that controls all aspects of spectral scanning and analysis.

FT-IR spectrometers are now marketed by a large number of manufacturers, and owing to competition among companies, the prices of spectrometers are progressively decreasing. At the same time, a wide variety of specialized sampling techniques are also becoming available, such as attenuated total reflectance (ATR), diffuse reflectance (DR), and FT-IR microscopy. Sophisticated software for infrared data analysis and manipulation is also available. Some of the methods that are particularly useful for the study of polypeptides and proteins include spectral subtraction, derivative, and deconvolution procedures.

3. Methods

3.1. Recording Infrared Spectra of Polypeptides and Proteins

Modern infrared spectrometers can be equipped with a wide range of sample holders and accessories to allow spectra to be obtained from a variety of sample forms. It is possible to obtain infrared spectra of proteins or polypeptides in different states, such as the solid state, thin films, and in solution. Since biological samples normally occur in an aqueous state, a brief description of how proteins and polypeptides can be studied in aqueous solution is given below.

The purified protein or polypeptide is dissolved in H_2O, or D_2O, or in any other suitable solvent as desired. For samples in D_2O, it is necessary to seal the sample container such that there is minimal contact with atmospheric water vapor. The sample concentration required may vary depending on the solvent used and also on the path length of the cell employed. In H_2O, a sample concentration of 10 mg/mL is normally sufficient for obtaining an FT-IR spectrum of good signal-to-noise ratio. In D_2O, a sample concentration as low as 1 mg/mL can be used. This is because of the lack of strong absorption arising from D_2O in the region 1700–1500 cm^{-1}. It is possible with D_2O to use cell path length of up to 100 μm, whereas with H_2O path lengths of only 6–12 μm can be used. The volume of sample required for one measurement may vary depending on the type of cell used, but in most cases, it may lie between 5 and 100 μL.

A variety of solution cells are commercially available for performing infrared measurements. As well as transmission cells, microcells with ATR optics can also be used to obtain spectra of proteins in aqueous solution *(20)*. However, we have found that in many cases proteins can become strongly adsorbed onto the surface of the ATR crystal producing erroneous results. We have found the use of transmission cells to be much simpler for most studies.

For our own studies, we use a Beckman FH-O1 CFT microcell. The choice of window material is dictated by the requirements for transparency in the mid-infrared region and minimal solubility of the material in water. Other important considerations are reflectance losses and relative cost. The material of choice in our case is calcium fluoride, which is almost insoluble in water and is transparent in the region 5000–1000 cm^{-1}. For studies down to 450 cm^{-1}, silver chloride may be used with the disadvantages, however, of both high reflection losses and high cost.

The microcell may be filled either by injection through one of the ports provided for this purpose or by placing about 10–50 μL of sample on the lower window, which supports the spacer, and carefully lowering (avoiding any air bubbles) the upper window to form a film of the sample. Injection of the sample is possible in those cases where the sample is available in sufficient quantities so that the dead volume of the cell is not a consideration. Spacers are commercially available, and are normally made of tin or Teflon™. The thickness of spacers can be 6–12 µm for samples in H_2O or up to 100 µm in the case of samples in D_2O. In the case of organic solvents that do not have absorbance near 1700–1500 cm^{-1}, it is possible to use a much greater path length. This enables good spectra to be obtained with a very dilute protein or polypeptide concentration. Once the sample is loaded, the cell can be fitted into a thermostatted cell holder. The sample then needs to be equilibrated at the chosen temperature for about 15 min. Furthermore, the sample compartment of the spectrometer needs to be purged with dry air or nitrogen to reduce the intense atmospheric water absorptions, which is particularly strong in the region 1700–1500 cm^{-1}. When band narrowing techniques, such as deconvolution or second derivative, are to be used, purging the sample compartment is particularly important. Once sufficient time (approx 15 min) is given for purging and temperature equilibration the sample is ready for scanning.

A different number of scans may be required to obtain spectra of satisfactory signal-to-noise ratio depending on the protein concentration used. The time required for scanning depends on the resolution and the type of spectrometer used. Thus, with Perkin-Elmer 1750 FT-IR spectrometers, 400 scans at a resolution of 4 cm^{-1} takes approx 1 h. The time taken doubles for the same number of scans if the resolution is increased to 2 cm^{-1}.

In order to obtain the spectrum of the protein or polypeptide, it is necessary to subtract the background absorption due to the buffer or solvent. Reference spectra need to be obtained of the solvent (with buffers if present) in which the protein or polypeptide is dissolved. The reference spectra must be obtained under exactly the same conditions (temperature, path length, number of scans, resolution, and so forth) as that used for the sample spectra. Precise matching of sample and buffer temperature is particularly important for samples in H_2O in order to avoid band shifts and the consequent generation of derivative shapes in the resulting absorption spectra.

Subtraction of D_2O absorbance from protein spectra is relatively straightforward, and is achieved by obtaining a straight baseline between 2000 and 1710 cm^{-1}. Subtraction of H_2O is more complicated because of its strong absorption near 1645 cm^{-1}. The spectra of protein and H_2O are displayed on the screen, and the contribution of H_2O is digitally subtracted via the interactive difference routine of the computer. In most cases, a good subtraction is obtained by cancellation of the H_2O combination band near 2150 cm^{-1}, where there is no absorbance from protein. This can be carried out by initially slightly oversubtracting and then decreasing the subtraction factor until a flat baseline is obtained between 2000 and 1710 cm^{-1}. It is noteworthy that spectra of proteins and polypeptides obtained in aqueous solution in different laboratories by different workers produce similar results. This has added confidence for the study of proteins in aqueous solution using FT-IR spectrsocopy. Figure 2 shows the spectrum of a soluble protein in H_2O before and after subtraction of the solvent absorbance.

With samples in D_2O, it is also possible to study the hydrogen-deuterium exchange of the amide groups by an examination of the amide II band *(21,22)*. The amide II band arises principally from an N–H bending vibration of the amide bond and the exchange of N–H to N–^{2}H leads to a shift of this band to a lower frequency by approx 100 cm^{-1}. It is thus possible to monitor the rate of hydrogen-deuterium exchange of a protein by following the change in intensity of the amide II band *(21–23)*. Deuteration of the amide bond also causes shift of the amide I band by few (approx 2–12 cm^{-1}) wave numbers. The magnitude of the shift of the amide I band has been used to distinguish between different secondary structures *(22,23)*. Figure 3 shows the absorption spectra of a soluble protein in H_2O and in D_2O after partial and complete hydrogen-deuterium exchange. Hydrogen-deuterium exchange studies have been found to be useful for detecting small conformational changes in various proteins *(22,24,25)*.

3.2. Data Analysis

After subtraction of the solvent absorbance, the spectrum of the protein is obtained and the amide I band can be analyzed using second-derivative and deconvolution procedures. From this the resulting amide I components can be assigned. Several references can be consulted regarding the assignment of amide I bands to different secondary structures *(5,6,11,12,14–16,22,23)*. Studies with model polypeptides and proteins of known secondary structure are leading to a greater clarification of the assignment of the amide I components. Bands in the region 1620–1640 cm^{-1} are normally assigned to β-sheet structure *(5,6,11,12,14–16,22,23)*. A second band associated with antiparallel β-sheet structure occurs in the region of 1670–1695 cm^{-1}. There is some disagreement in

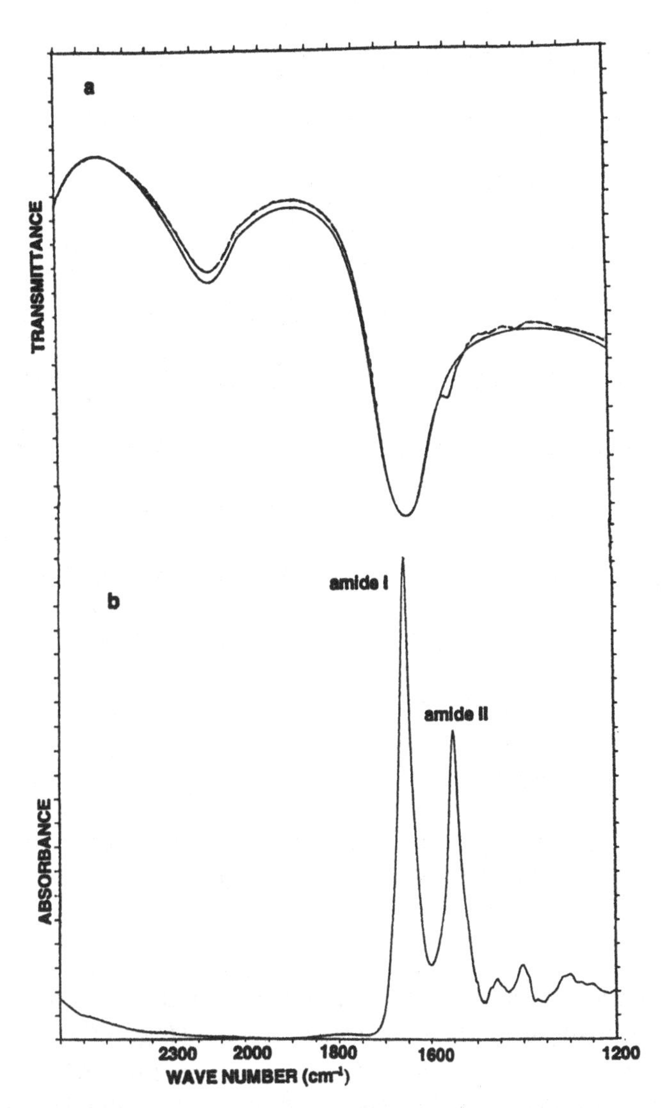

Fig. 2. (**a**) FT-IR transmittance spectrum of H_2O (continuous line) and of the soluble protein myoglobin dissolved in H_2O (broken line). The protein concentration used was 50 mg/mL. (**b**) The absorbance spectrum of myoglobin generated from Fig. 2a after the subtraction of the H_2O spectrum from the spectrum of myoglobin dissolved in H_2O.

the literature as to which one of the several components observed in this region can be associated with β-sheet *(14,22,23)*, since β-turn structure also absorbs in this region. It has been reported that 3_{10}-helical structure can absorb near 1665 cm^{-1} *(13,16,26)*.

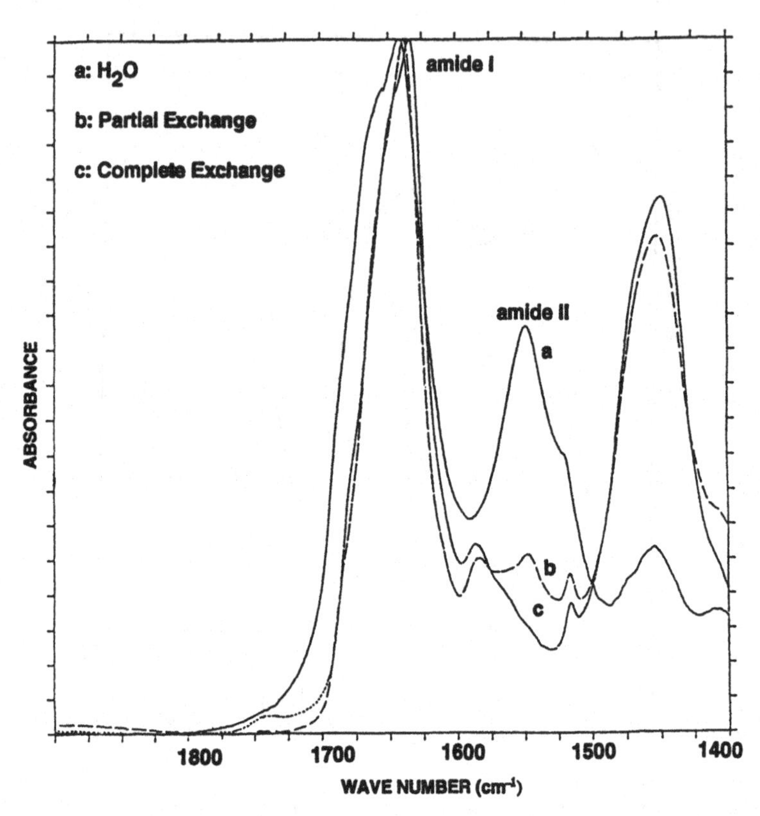

Fig. 3. Absorbance spectrum of the soluble protein ribonuclease A in H_2O (**a**) in D_2O after partial hydrogen-deuterium exchange (**b**) and after complete hydrogen-deuterium exchange in D_2O (**c**).

The α-helical structure present in proteins and polypeptides in H_2O and D_2O normally absorbs in the range of 1648–1657 cm^{-1}. With proteins and polypeptides in H_2O, there can be an overlap of absorption arising from α-helical and unordered structure *(11,12)*. However, these two secondary structures can be distinguished after carrying out measurements in D_2O. The absorption of unordered structure occurs at a lower band frequency as a result of hydrogen-deuterium exchange of its amide N–H groups. Thus, in D_2O, unordered structure absorbance occurs at approx 1644 cm^{-1}, whereas α-helical structure absorbs at approx 1648–1657 cm^{-1} *(5,6,11,12,14–16,22,23)*.

Some typical FT-IR studies of protein structures are shown in Figs. 4 and 5. Figure 4A shows the absorbance spectrum (after digital subtrac-

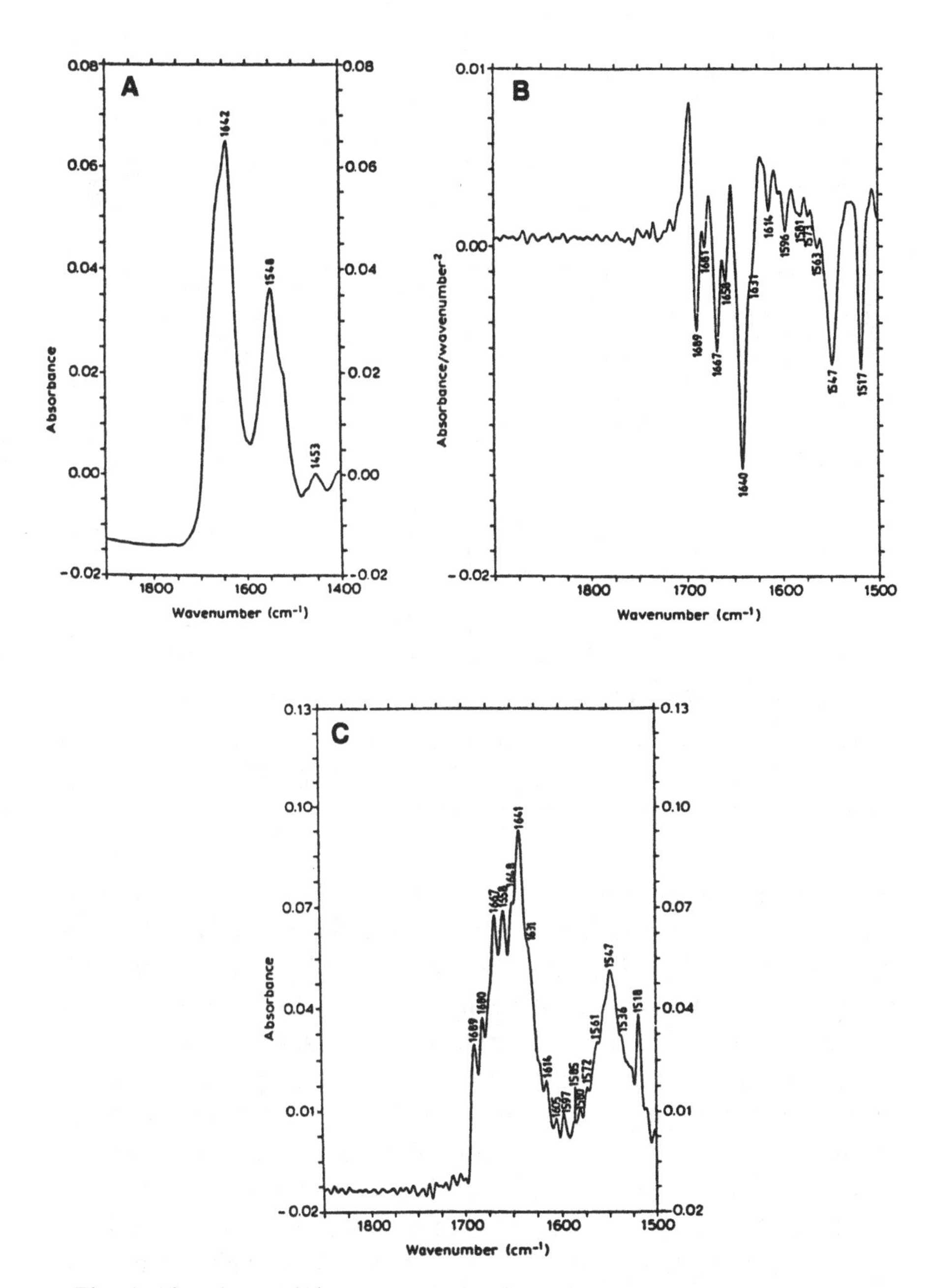

Fig. 4. Absorbance (**A**), second-derivative (**B**), and deconvoluted (**C**) spectra of ribonuclease A in H_2O phosphate buffer (*see* Haris et al. *[22]*).

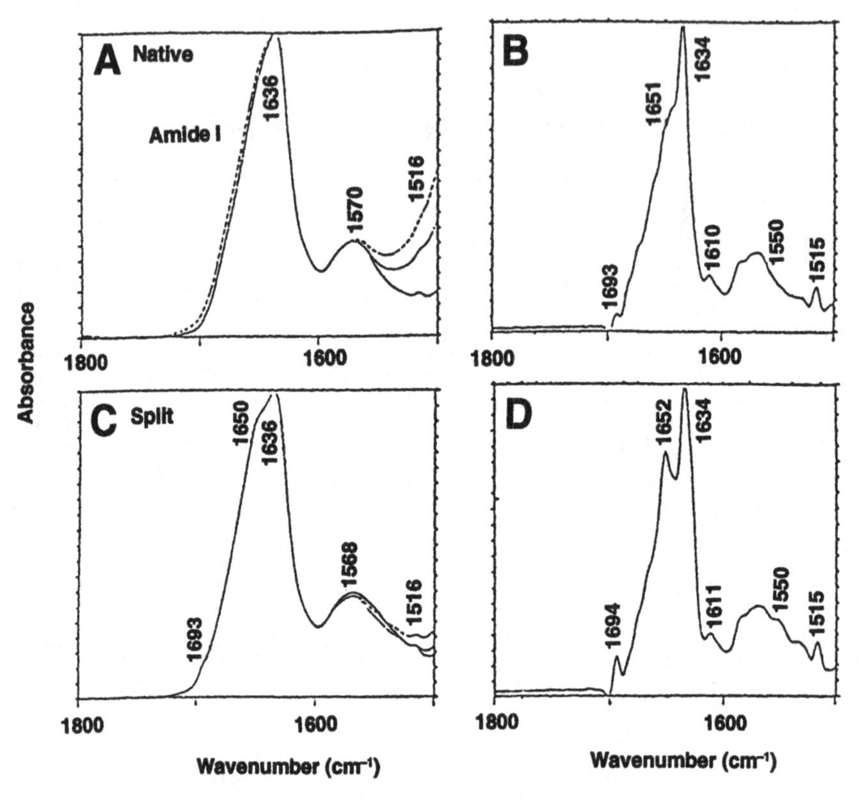

Fig. 5. FT-IR spectra of native and split α_1-antitrypsin. Panels **A** and **C** correspond to the absorbance spectra whereas panels **B** and **D** correspond to the deconvoluted spectra. For further details, *see* Haris et al. *(25)*.

tion of the buffer spectrum) of ribonuclease A in H_2O buffer. Two main bands are seen in the spectral region shown: the amide I band at 1642 cm^{-1} and the amide II band at 1548 cm^{-1}. These bands are, themselves, composed of a number of absorptions as revealed by the second-derivative and deconvoluted spectra presented in Figs. 4B and 4C, respectively. Minima in the second derivative correspond to positive absorptions in the original absorbance spectrum. The deconvoluted and the second-derivative spectra show similar bands in the amide I region. One additional feature at 1647 cm^{-1} is observed in the deconvoluted spectrum. The 3D structure of ribonuclease A determined using X-ray diffraction is known to be predominantly β-sheet with a substantial amount of α-helices *(27)*. This agrees with the FT-IR spectroscopic data, where a major band corresponding to β-sheet in both the second-derivative and deconvoluted spectra can be seen at 1639 cm^{-1}. The presence of a substantial amount of α-helical structure is indicated by a band at 1657 cm^{-1}. The most accurate assignment of amide bands is possible when spectra are recorded in both H_2O and D_2O. This has been reported for ribonuclease A and S *(22,23)*. Deuteration of the amide groups leads to a low-frequency shift of their infrared bands. The magnitude of the shift on replacement of H_2O with D_2O reflects the pattern of hydrogen bonding associated with the structures representing these bands. This is found to be useful in distinguishing between different secondary structures in ribonuclease A and S, and for making accurate assignments *(22,23)*.

Hydrogen-deuterium exchange studies using infrared spectroscopy has been found to be useful for detecting small conformational differences among proteins. For example, the degree of hydrogen-deuterium exchange is markedly less for ribonuclease A than for ribonuclease S *(22)*. These two proteins have similar enzymatic activity and only differ by the presence of a cleavage of the peptide bond between amino acids 20–21 in ribonuclease S. The greater degree of hydrogen-deuterium exchange for ribonuclease S can be explained by the fact that the cleavage of the peptide bond in this protein produces a less rigid structure that has greater accessibility to D_2O *(22)*. Further information regarding assignments of bands in the spectra of ribonuclease A and S can be found in Haris et al. 1986 *(22)*.

Figure 5 shows the FT-IR spectra of α_1-antitrypsin *(25)* recorded in D_2O. α_1-Antitrypsin is the best-characterized member of the SERPIN superfamily of plasma proteins *(28)*. Protease inhibitors of this family undergo a characteristic reactive-center cleavage during expression of their inhibitory activity. The physical basis of this transition in α_1-antitrypsin from the stressed native conformation to the more stable reactive center cleaved (split) form was studied by FT-IR spectroscopy *(25)*. The amide I band in both cases is centered at 1636 cm^{-1}. The feature at 1568–1570 cm^{-1} corresponds to absorption from carboxylate side chains *(29)*. The 1516-cm^{-1} component is normally attributed to tyrosine side chain absorption *(30)*. The frequency of the main amide I band for both samples is consistent with the presence of a predominantly β-sheet structure in α_1-antitrypsin *(31)*. However, native and split α_1-antitrypsin differ in that additional features near 1650 and 1693 cm^{-1} are present in the split form. These were reproducibly observed in 0.05, 0.50, and 1.25*M* phosphate *(25)*. The deconvoluted spectra of the two forms in panels B and D of Fig. 5 emphasize the differences. The main β-sheet feature at 1634 cm^{-1} is visible in both forms. The split form shows considerably more intense β-sheet and α-helix bands at 1694 and 1652 cm^{-1}. The bands near 1650 and 1690 cm^{-1} have been attributed to α-helix and antiparallel β-sheet structures, respectively *(22,23)*. The FT-IR spectroscopic results clearly show that the α-helical and antiparallel β-sheet content have significantly increased after reactive-center cleavage of α_1-antitrypsin. In the deconvoluted spectra shown in Fig. 5, a peak at 1550 cm^{-1} corresponding to the amide II band is visible for split α_1-antitrypsin, but not for native α_1-antitrypsin. These indicate that hydrogen-deuterium exchange is more extensive in the native form, as might be expected from its more open structure *(25)*.

In most cases, a good relationship exists between the major protein secondary structures and their respective amide I frequencies. However, complexities can occur with proteins containing unusual or less common structures. It is also important to note that absorption from amino acid side chains can also occur in the amide I region *(32)*. The investigator must continue to exercise due caution when assigning bands in the amide I region. The amide II band has not been as well studied as the amide I band, and the frequency of the amide II band is only poorly correlated with different protein secondary structure. However, as discussed, the amide II band is useful for studying hydrogen-deuterium exchange of the peptide groups.

The advantages as well as the potential pitfalls of deconvolution and derivative techniques as methods for band narrowing have been discussed by Mantsch et al. *(18)*. Problems associated with these procedures includes the appearance of side lobes and the fact that sharp water vapor features and random noise are disproportionately enhanced in the deconvoluted and second-derivative spectra. Thus, it is essential to have absorption spectra of very high signal-to-noise ratio in order to apply these enhancement procedures. It is common practice to apply both deconvolution and second-derivative methods to a spectrum, and assign peaks seen using both of these methods. Furthermore, it may be useful to subtract a spectrum of water vapor away from the protein spectrum.

There are a large number of studies reported in the literature where FT-IR spectroscopy has been applied to study the changes in protein and polypeptide structure as a result of variation of pH, solvent composition, temperature, light, ligand binding, exposure to lipids, and so on (for recent reviews, *see* refs. *5,6,15,16*). However, most of these studies are of a qualitative nature. Quantitative analysis can provide a greater understanding of the nature and extent of the changes involved. Some quantitative studies have been reported *(14,16)*. The method commonly used is based on identifying the number of amide I component bands and their approximate positions from deconvoluted spectra, and then curve fitting the overall amide I band contours. Although good results have been obtained, this method is to some extent subjective. The difficulties faced with this approach have been discussed *(5)*. One of the major problems associated with this method is the need to assume a band shape for each of the components. It is usually assumed that each component has the same band shape.

We have recently reported an alternative method to curve fitting *(33)*. We used a factor analysis approach to analyze the infrared spectra of 18 soluble proteins whose crystal structures are known from X-ray studies. The analysis is performed by using the program CIRCOM (Computerized Infrared Characterization of Materials) available from Perkin-Elmer (Norwalk, CT). A full mathematical description of this method has been given *(34)*. Initially, a calibration set is generated from the infrared spectra of a range of soluble proteins of known X-ray crystallographic structure. Factor analysis followed by multiple linear regression identifies those eigenspectra that correlate with the variation in properties described by the calibration set. The properties of interest in this study are % α-helix, % β-sheet, and % turns. In the analysis of an unknown, the factor loadings required to produce its spectrum are substituted in the regression equation for each property to predict its secondary structural composition. The accuracy of the method is determined by removing each standard, in turn, from the calibration set and using a calibration set generated from the remainder to predict its composition. By this method, we obtain standard error of predictions of 3.9% for α-helix, 8.3% for β-sheet, and 6.6% for turns. From Table 2, it can be seen that in most cases agreement between the CIRCOM and X-ray diffraction values for the three classifications of secondary structure is good.

This method has important advantages for the analysis of protein structure when compared to the commonly used method of deconvo-

Table 2
Prediction of Protein Secondary Structure Using CIRCOM[a]

Protein	Helix C[b]	Helix X[c]	Sheet C	Sheet X	Turns C	Turns X
Myoglobin	86	88	–8	0	15	7
Hemoglobin	81	86	16	0	12	8
Insulin	53	61	21	15	23	12
Cytochrome C	53	49	21	11	15	22
Lysozyme	55	46	5	19	13	23
Alcohol dehyrogenase	30	29	44	40	20	19
Papain	26	28	32	29	21	18
Nuclease	26	26	36	37	26	23
Ribonuclease A	25	23	38	46	16	21
Ribonuclease S	23	23	38	53	24	15
Carbonic anhydrase	12	16	52	45	24	25
Chymotrypsinogen	16	12	46	49	21	23
Protease	15	11	50	57	16	18
Chymotrypsin	11	11	50	50	22	25
Elastase	9	10	55	46	18	28
Prealbumin	2	6	61	61	28	19
Concanavanlin	2	3	70	65	30	22

[a]For further details, *see* ref. *33*.
[b]C = CIRCOM predictions.
[c]X = X-ray values.

lution followed by curve fitting *(14,16)*. First, no deconvolution (or generation of derivatives) is required, and pretreatment of the data is kept to a minimum. It is known that overdeconvolution produces negative side lobes on absorption bands, and can result in artificial bands from noise or incomplete compensation of water vapor. Thus, several distortions may be induced in the spectrum before curve fitting is applied. Second, no assignment of the amide I component is necessary. This is an important feature, since there are several instances where an assignment of all the bands is impossible to make with complete certainty. In particular, as mentioned earlier on, it is difficult to differentiate the high wave-number component associated with antiparallel β-sheet from absorptions associated with β-turns. A difficulty with the deconvolution/curve-fitting approach lies in the inaccuracies in the curve-fitting step. Typically, only those components observed by deconvolution are used to generate the band shape of either the original or deconvoluted *(14)* spectrum. In our experience, this results in an incomplete fit and a nonunique solution, casting doubt on the accuracy of the quantitative determination derived from the band areas of the components. A better fit can usually be obtained by adding weaker components not revealed by deconvolution. Although it is certain that neither derivative or deconvoluted spectra reveal all the component bands, the user may have little justification in accepting and assigning these extra bands.

One assumption common to curve-fitting methods and our new method is that the extinction coefficients of the bands assigned to α-helical, β-sheet, and turn structures are identical. There is some evidence that changes in extinction coefficients may occur *(35)*. It appears however, that variations in path length are more important than variations in absorption coefficient. Although the predictions generated by the CIRCOM method are good, there still remains scope for improvement. An increase in the number of standards of known structure in the calibration set should improve the accuracy of the prediction of an unknown. A single large calibration set or sets specific to narrow ranges of secondary structure could be used. A combination of CD spectroscopy and FT-IR spectroscopy may prove valuable for such quantitative studies.

FT-IR spectroscopy is a useful technique for obtaining information about the secondary structure content of polypeptides and proteins in their native aqueous environment. It has important advantages over other techniques commonly used in the study of proteins and polypeptides. This is particularly the case with membrane-associated polypeptides and proteins.

A recent development is the introduction of isotopic labels at specific sites of peptides or proteins. The most useful appears to be the substitution of ^{13}C for ^{12}C and ^{15}N for ^{14}N. This approach was used for the study of protein–protein interaction *(36)*. The advantages and drawbacks of FT-IR spectroscopy for structural analysis of proteins has been discussed in a recent article (*see* ref. *37*).

Acknowledgments

We wish to thank the Wellcome Trust for financial support.

References

1. Elliot, A. and Ambrose, E. J. (1950) Structure of synthetic polypeptides. *Nature* **165,** 921–922.
2. Susi, H. (1969) Infrared spectra of biological macromolecules and related systems, in *Structure and Stability of Biological Macromolecules* (Timasheff, S. N. and Fasman, G. D., eds.), Marcel Dekker, New York, pp. 575–663.
3. Cameron, D. G., Casal, H. L., and Mantsch, H. H. (1979) Application of Fourier transform infrared transmission spectroscopy to the study of model and natural membranes. *J. Biochem. Biophys. Meth.* **1,** 21–36.
4. Chapman, D., Gomez-Fernandez, J. C., Goni, F. M., and Barnard, M. (1980) Difference infrared spectroscopy of aqueous model and biological membranes using an infrared data station. *J. Biochem. Biophys. Meth.* 2, 315-323.
5. Jackson, M., Haris, P. I., and Chapman, D. (1989) Fourier transform infrared spectroscopic studies of lipids, polypeptides and proteins. *J. Mol. Struc.* **204,** 329–355
6. Braiman, M. S. and Rothschild, K. J. (1988) Fourier transform infrared techniques for probing membrane protein structure. *Ann. Rev. Biophys. Chem.* **17,** 541–570.
7. Susi, H. (1972) Infrared spectroscopy—conformation. *Methods Enzymol.* **26,** 455–472.
8. Miyazawa, T., Shimanouchi, T., and Mizushima, S. I. (1956) Characteristic infrared bands of mono-substituted amides. *J. Chem. Phys.* **24,** 408–418.
9. Anderle, G. and Mendelshon, R. (1987) Thermal denaturation of globular proteins—Fourier transform infrared studies of the amide III spectral region. *Biophys. J.* **52,** 69–74.
10. Chirgadze, Y. N., Brazhnikov, E. V., and Nevskaya, N. A. (1976) Intramolecular distortion of α-helical structure of polypeptides. *J. Mol. Biol.* **102,** 781–792.
11. Susi, H., Timasheff, S. N., and Stevens, L. (1967) Infrared spectra and protein conformations in aqueous solutions. *J. Biol. Chem.* **242,** 5460–5466.
12. Timasheff, S. N., Susi, H., and Stevens, L. (1967) Infrared spectra and protein conformations in aqueous solutions. *J. Biol. Chem.* **242,** 5467–5473.
13. Krimm, S. and Bandekar, J. (1986) Vibrational spectroscopy and conformation of peptides, polypeptides and proteins. *Adv. Protein Chem.* **38,** 181–364.
14. Byler, D. M. and Susi, H. (1986) Examination of the secondary structure of proteins by deconvoluted FTIR spectra. *Biopolymers* **25,** 469–487.
15. Susi, H. and Byler, D. M. (1986) Resolution-enhanced Fourier transform infrared spectroscopy of enzymes. *Methods Enzymol.* **130,** 290–311.
16. Surewicz, W. K. and Mantsch, H. H. (1988) New insight into protein secondary structure from resolution-enhanced infrared spectra. *Biochem. Biophys. Acta* **952,** 115–130.
17. Kauppinen, J. K., Moffatt, D. J., Mantsch, H. H., and Cameron, D. G. (1981) Fourier self deconvolution—a method for resolving intrinsically overlapped bands. *Appl. Spectrosc.* **35,** 271–276.
18. Mantsch, H. H., Casal, H. L., and Jones, R. N. (1986) Resolution enhancement of infrared spectra of biological systems, in *Spectroscopy of Biological Systems* (Clark, R. J. H. and Hester, R. E. eds.), Wiley, New York, pp. 1–46.

19. Griffiths, P. R. and de Haseth, J. A. (1986) *Fourier Transform Infrared Spectrometry*, Wiley, New York.
20. Dev, S. B., Keller, J. T., and Rha, C. K. (1988) Secondary structure of 11 S globulin in aqueous solution investigated by FT-IR derivative spectroscopy. *Biochem. Biophys. Acta* **957,** 272–280.
21. Barksdale, A. D. and Rosenberg, A. (1982) Acquisition and interpretation of hydrogen exchange data from peptides, polymers, and proteins. *Meth. Biochem. Anal.* **28,** 1–113.
22. Haris, P. I., Lee, D. C., and Chapman, D. (1986) A Fourier transform infrared investigation of the structural differences between ribonuclease A and ribonuclease S. *Biochem. Biophys. Acta* **874,** 255–265.
23. Olinger, J. M., Hill, D. M., Jakobsen, R. J., and Brody R. S. (1986) Fourier transform infrared studies of ribonuclease in H_2O and H_2O solutions. *Biochem. Biophys. Acta* **869,** 89–98.
24. Alvarez, J., Haris, P. I., Lee, D. C., and Chapman, D. (1987) Conformational changes in concanavalin A associated with demetallization and α-methylmannose binding studied by Fourier transform infrared spectroscopy. *Biochem. Biophys. Acta* **916,** 5–12.
25. Haris, P. I., Chapman, D., Harrison, R. A., Smith, K. F., and Perkins, S. J. (1990) Conformational transition between native and reactive center cleaved forms of α_1-antitrypsin by Fourier transform infrared spectroscopy and small-angle neutron scattering. *Biochemistry* **29,** 1377–1380.
26. Haris, P. I. and Chapman, D. (1988) Fourier transform infrared spectra of the polypeptide alamethicin and a possible structural similarity with bacteriorhodopsin. *Biochem. Biophys. Acta* **943,** 375–380.
27. Wlodawer, A., Bott, R., and Sjolin, L. (1982) The refined crystal structure of ribonuclease A at 2.0 A resolution. *J. Biol. Chem.* **257,** 1325–1332.
28. Carell, R. and Travis, J. (1985) α_1-antitrypsin and the serpins—variation and conservation. *Trends Biochem. Sci.* **10,** 20–24.
29. Holloway, P. W. and Mantsch, H. H. (1988) Infrared spectroscopic analysis of salt bridge formation between cytochrome b_5 and cytochrome c. *Biochemistry* **27,** 7991–7993.
30. Susi, H. and Byler, D. M. (1983) Protein structure by Fourier transform infrared spectroscopy : second derivative spectra. *Biochem. Biophys. Res. Commun.* **115,** 391–397.
31. Lobermann, H., Tokuoka, R., Deisenhofer, J., and Huber, R. (1984) Human α_1-Proteinase inhibitor—crystal structure analysis of two crystal modifications, molecular model and preliminary analysis of the implications for function. *J. Mol. Biol.* **177,** 531–556.
32. Chirgadze, Yu. N., Fedorov, O. V., and Trushina, N. P. (1975) Estimation of amino acid residue side-chain absorption in infrared spectra of protein solutions in heavy water. *Biopolymers* **14,** 679–694.
33. Lee, D. C., Haris, P. I., Chapman, D., and Mitchell, R. C. (1990) Determination of protein secondary structure using factor analysis of infrared spectra. *Biochemistry* **29,** 9185–9193.

34. Malinowski, E. R. and Howery, D. G. (1980) *Factor Analysis in Chemistry*, Wiley, New York.
35. Jackson, M., Haris, P. I., and Chapman, D. (1989) Conformational transitions in poly(L-lysine): studies using Fourier transform infrared spectroscopy. *Biochem. Biophys. Acta* **998,** 75–79.
36. Haris, P. I., Robillard, G. T., van Dijk, A. A., and Chapman, D. (1992) Potential of ^{13}C and ^{15}N labeling for studying protein–protein interactions using Fourier transform infrared spectroscopy. *Biochemistry* **31,** 6279–6284.
37. Surewicz, W. K., Mantsch, H. H., and Chapman, D. (1993) Determination of protein secondary structure by Fourier transform infrared spectroscopy: a critical assessment. *Biochemistry* **32,** 389–394.

CHAPTER 15

Fluorescence Spectroscopy

Paul G. Varley

1. Introduction

1.1. Mechanisms of Fluorescence

When a substance absorbs light, its molecules are excited from a ground state to a higher energy level. The molecule can then return to the ground state by losing energy, mainly as heat to its surroundings. In some cases, part of the energy absorbed can be reemitted as radiation, usually of a longer wavelength than the exciting light. This process is known as fluorescence *(1)*.

How the light is absorbed and emitted during the fluorescence process is shown in Fig. 1. The electronic states of the system are represented as A, the ground state, and B, the first excited state. Each electronic state can exist in different vibrational energy levels shown here by lines 0, 1, 2, and so on.

Following the absorption of light, electronic excitation occurs usually to a higher vibrational level of B. At this stage if the substance is not fluorescent, the energy will be lost as heat as the molecule returns to the ground state. If, however, the molecule is fluorescent (i.e., contains a fluorophore), then some energy will be lost allowing the molecule to reach the lowest vibrational level of the excited state. Return to the ground state would then be achieved by reradiation of the absorbed energy as fluorescence.

Fluorescence emission has various characteristics of which there are usually few exceptions:

From: *Methods in Molecular Biology, Vol. 22: Microscopy, Optical Spectroscopy, and Macroscopic Techniques* Edited by: C. Jones, B. Mulloy, and A. H. Thomas

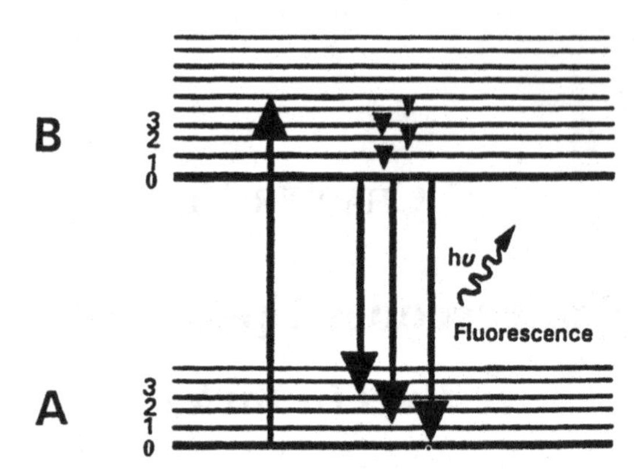

Fig. 1. The absorption and emission of light during the fluorescence process. The molecule in the lowest vibrational level (0) of the ground state (**A**) absorbs a quantum of light, and an electron is promoted to a higher energy orbital with a concomitant change in the vibrational state of the molecule. The vibrationally excited state of the electronically excited state (**B**) loses vibrational energy by thermal equilibriation with the solvent. An electronic transition to a higher vibrational state of the ground state may occur with release of a quantum of light. This is fluorescence.

1. Since energy is lost during the drop to the lowest vibrational level of the excited state, the fluorescent radiation will have less energy than the exciting light. This will result in the fluorescence having a lower frequency and hence longer wavelength than the exciting light (note that the energy will be further reduced, since fluorophores usually decay to excited vibrational levels of the electronic ground state; *see* Fig. 1).
2. Fluorescence spectra are generally mirror images of their *absorbance* spectra. Any fine structure observed in *absorption* spectra reflects the vibrational energy levels of the excited state. Similar structures observed in the fluorescence spectra reflect the vibrational levels in the ground state. Since the spacings of the these levels in the ground and excited states are similar, vibrational structure seen in the absorption spectra would also be seen in the fluorescence emission.
3. Fluorescence emission spectra are generally independent of the wavelength of the exciting light. Since the time taken for the fluorophore to drop down to the lowest vibrational level in the excited state is short compared to the total time spent in the excited state, the lowest level will always be reached before fluorescence occurs. This means that the resulting spectra will depend on the probabilities of falling into the ground state only. The wavelength of excitation, which defines the vibrational level reached in the excited state, will therefore not affect the wavelength of the fluorescence.

1.2. Fluorescent Compounds

The types of substance that normally exhibit fluorescence are those whose structure contains delocalized electrons present in conjugated double bonds. The fluorescent compounds of interest in biochemistry and molecular biology can be divided into intrinsic and extrinsic fluorophores. Intrinsic or naturally occurring fluorophores include tryptophan, tyrosine, and cofactors, such as NADH and some rare bases in nucleic acids, e.g., the Y-base in yeast Phe-tRNA. Extrinsic fluorophores are a wide and varied range of compounds including protein labels, e.g., fluorescein, membrane probes, such as 1,6-diphenylhexatriene (DPH), and DNA-binding compounds, such as ethidium bromide.

1.3. Why Measure Fluorescence?

Fluorescence has the main advantage over other optical techniques of being extremely sensitive. Many applications of fluorescence rely on this fact to label small amounts of sample for detection, often allowing solute in the nanomolar range to be quantified.

The fluorescence of many fluorophores is extremely sensitive to their surrounding environments, e.g., tryptophan in proteins. This enables fluorescence to be used as a sensitive structural probe in many biological systems using only small amounts of sample compared to other spectroscopic techniques.

2. Fluorescence Measurements

2.1. Fluorescence Parameters

There are many parameters that can be measured to characterize the fluorescence, depending on the type of information required.

1. The most basic fluorescence experiments irradiate the sample with light at or close to its absorption maximum, and measure the resulting fluorescence at a particular emission wavelength or as a function of emission wavelength (emission spectra). Emission spectra are also often presented with an excitation spectrum where the fluorescence at a single wavelength is measured as a function of excitation wavelength. Any fluorescence process can be characterized by a quantum yield, which is the ratio of the intensity of the fluorescent light to the exciting light, i.e., the efficiency of the fluorescence process. This is generally measured by comparing the sample under study with a compound of known quantum yield at a similar wavelength. These basic fluorescence measure-

ments, which use a standard commercially available spectrometer, are often known as steady-state fluorescence measurements to distinguish them from fluorescence lifetimes (*see* Section 2.2.).

2. Fluorescence lifetime is the average time the fluorophore spends in the excited state. Lifetimes are generally of the order of 10^{-8} s (10 ns) and require specialized equipment in order to be measured. Since the functionally important fluctuations that occur in biological systems and fluorescence lifetimes occur on a similar time scale, measurement of these lifetimes is a useful probe of the dynamics in these systems.
3. Fluorescence anisotropy or polarization is the degree to which fluorescence emission is depolarized relative to polarized exciting light. The absorption of light by molecules depends on the relative orientation of the exciting photon's electric vector and the transition moment of the fluorophore. If the fluorophore is excited by polarized light, then only molecules whose transition moment is in the correct orientation (i.e., parallel to the photon's electric vector) will be excited, and the resulting fluorescence will be partially polarized. The extent of depolarization depends on the movement of the fluorophore during the fluorescence lifetime. The measurement of such parameters can therefore be used to measure the degree of rotational freedom of the fluorophore on the lifetime time scale.

   The depolarization can be quantified by either polarization or anisotropy:

$$\text{Polarization} = (I_{\parallel} - I_{\perp}/I_{\parallel} + I_{\perp}) \quad (1)$$

$$\text{Anisotropy} = (I_{\parallel} - I_{\perp}/I_{\parallel} + 2I_{\perp}) \quad (2)$$

   where $I_{\parallel}$ and $I_{\perp}$ are the fluorescence intensities of vertically and horizontally polarized light, respectively, when the sample is excited with vertically polarized light. Anisotropy is the generally preferred property to use, since it is simpler than polarization when the majority of theoretical expressions are considered.

   Measurements are made by exciting with vertically polarized light and measuring the resulting fluorescence through polarizers in the vertical and horizontal orientations. Fluorescence intensity (steady state) or fluorescence lifetimes can be measured in such studies. When the lifetimes are determined, results are normally presented as a function of time, to give the change in anisotropy as the fluorescence decays. For steady-state measurements, anisotropy is normally presented as a function of another parameter, e.g., temperature, viscosity, wavelength, and so forth.
4. Fluorescence quenching is the means by which some compounds have the ability to decrease the fluorescence of some fluorophores *(2)*. There

are two kinds of quenching: static, which involves a complex formation between the quencher and the fluorophore, and dynamic quenching, which requires a collision between the two substances. In each case, contact between the fluorophore and quencher during the lifetime of the excited state causes the fluorophore to return to the ground state, transferring its energy to the quencher without fluorescence occurring. The result is a decrease in fluorescence intensity and lifetime. Since contact between fluorophore and quencher is required, the ability to quench a particular fluorescent group is indicative of the accessibility and dynamics of the system under study. Quenching experiments are performed by titrating quencher into the sample under study, and measuring the decrease in intensity or lifetimes. Data is normally analyzed using the Stern-Volmer equation:

$$(F_0/F) = (\tau_0/\tau) = 1 + k_q \tau_0[Q] \quad (3)$$

where F_0 and F are the fluorescence intensity in the absence and presence of quencher, Q, at a concentration $[Q]$, τ_0 and τ are the fluorescence lifetimes in the absence and presence of quencher, and k_q is the quenching rate or bimolecular rate constant. The product $k_q\tau_0$ or the Stern-Volmer quenching constant is determined as the gradient of a plot of F_0/F or τ_0/τ against $[Q]$. Note that for determination of the bimolecular rate constant, at least the lifetime in the absence of quencher needs to be determined. This analysis applies to dynamic quenching only. For static quenching, which is characterized by upwardly curving Stern-Volmer plots, the analysis is more complex.

5. Energy transfer is the effect where an electronically excited fluorophore (donor) can transfer its energy to another fluorophore (acceptor), which then emits the energy as fluorescence *(3)*. The amount of energy transfer depends on the extent of overlap of the donor's fluorescence and the acceptor's absorption spectra, since the more they overlap, the more efficient transfer will be. The efficiency of the process also depends on the relative orientation of the donor's and acceptor's transition dipoles, and the distance they are apart. The major application of energy transfer is therefore as a measure of distance on the atomic scale between fluorescent groups. After determination of the quantum yield of the fluorescence of the acceptor in the presence (Φ_T) and absence (Φ_D) of the donor (or the lifetime of the acceptor in the presence and absence of the donor) the transfer efficiency (E_T) can be determined. Using R_0, the distance of 50% transfer efficiency (which is a function of the overlap between the donor's emission and the acceptor's absorption spectra), the distance between the two groups, R, can be determined thus:

$$1 - E_T = (\Phi_T/\Phi_D) \quad (4)$$

and

$$R = R_0\,(1 - E_T/E_T)^{1/6} \quad (5)$$

2.2. Instrumentation

There are two basic types of spectrofluorometers: those that measure steady-state fluorescence and those that measure fluorescence lifetimes. The basic design of each instrument is, however, the same and is shown in Fig. 2.

The light source passes through a monochromator (or filter), which selects the excitation wavelength. This is focused on the cell containing the sample. The fluorescence is detected perpendicular to the excitation direction to avoid detection of light passing straight through the sample. The fluorescence is focused onto a monochromator, which selects the desired wavelength before quantitation by a photomultiplier tube (PMT). Variable slits on both the excitation and emission sides allow the intensity and bandwidth of the light to be controlled. For anisotropy measurements, the polarizers are placed between the focusing lenses on and the monochromators on both the emission and excitation paths.

2.2.1. Steady-State Spectrofluorometers

Steady-state fluorometers simply measure the intensity of the fluorescence at the desired wavelengths. Most machines use xenon lamps that provide relatively continuous light over the region applicable to most fluorescence applications. Monochromators are generally of the grating type. There are two kinds of detection used in these fluorometers, single-photon counting and analog measurement. In single-photon counting machines, each photon is registered as a pulse on the PMT, whereas in an analog detection system, the pulses are averaged. The single-photon counting is more sensitive (since it operates on the theoretically maximum sensitivity) and generally found on the more expensive machines. It does, however, have some drawbacks, including a limited intensity range that results from the need to avoid two or more photons arriving at the detector simultaneously, since this would result in detection of only one pulse. The less sensitive analog system has advantages over its single-photon counting counterpart in general robustness and the facility to measure higher intensities of fluorescence.

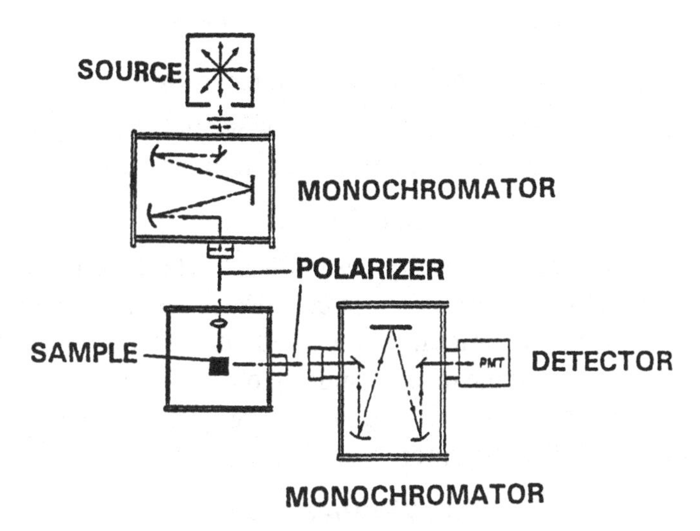

Fig. 2. Schematic layout of a typical fluorometer. A collimated excitation beam of defined wavelength is produced by the source and monochromator 1, and excites the sample. Fluorescence is measured perpendicular to the exciting beam after passing through the second monochromator.

2.2.2. Time-Resolved Spectrofluorometers

The measurement of fluorescence lifetimes is a more complex matter than that of measuring steady fluorescence. The two main methods for measuring lifetimes are the phase-modulation method and the pulse method, which is usually used with single-photon counting.

In the pulse method, the sample is excited by light whose intensity is sinusoidally modulated. The phase change and demodulation of the fluorescence when compared to the exciting light can then be used to determine the lifetimes. In the pulse method, the sample is excited by short pulses of light, and the decay of the resulting fluorescence is measured. This decay can then be analyzed to obtain the lifetimes. Each technique has its own advantages. The phase method is fast and good at measuring short (subnanosecond) lifetimes. The pulse method is more suited to the resolution of heterogeneous fluorescence decays and is more sensitive than the phase method, which means less sample is generally required. Recently, the resolution of time-resolved fluorescence measurements has been improved by using improved light sources, principally those from laser or synchrotron radiation.

3. Practical Considerations

Various practical considerations *(4,5)* when performing any kind of fluorescence experiment will ensure accuracy and consistency.

3.1. Cuvet

Fluorescence cuvets usually resemble standard 1-cm absorption cuvets, except that all sides are optical surfaces to allow detection of the fluorescence perpendicular to the incident light. Care should therefore be taken to avoid the direct handling of the optical surfaces. Also when cleaning the cuvet with detergent, be sure to rinse well, since many detergents contain fluorescent substances.

3.2. Temperature

Fluorescence, particularly intrinsic protein fluorescence, is extremely sensitive to temperature. Fluorescence experiments should therefore always be carried out with the cell holder connected to a thermostatically controlled circulator, which can accurately maintain the desired temperature.

3.3. Sample

Probably the most important consideration in the preparation of a sample for fluorescence is its concentration. The sample should have an absorbance at the excitation wavelength of <0.1. This is because of the inner filter effect. Light absorbed at the front of a sample will be absorbed so that the sample further on in the pathlength of the cuvet will "see" less light. This effect increases greatly with increasing absorbance of the sample, and can decrease the intensity of fluorescence by either lowering the effective intensity of the exciting light or absorbing some of the fluorescence. In either case, the fluorescence intensity, which should be proportional to the absorbance, will not be so with the discrepancy increasing as the sample's absorbance increases. Samples of low absorbance (typically 0.01–0.05) should be used. Alternatively, if this is not possible, calibration curves of fluorescence against absorbance may be used to quantify it for a particular system, allowing the correct adjustments to be made.

Where possible, samples should be free of all types of turbidity, e.g., dust, aggregates, and so forth, since this promotes light scattering, which can affect the measurement by either filtering out the exciting light or causing the scatter peak at the excitation wavelength to overlap with the region of interest. Long-term exposure of samples, buffers, and so on, to plastics is not recommended to avoid possible contamination by fluorescent substances often present in such materials. Samples are often bubbled through with nitrogen before measurement to eliminate oxygen from the solution, which prevents quenching by dissolved oxygen of certain fluorophores.

3.4. Baseline Measurement

When recording fluorescence spectra, a baseline should always be measured. This ensures that the solution containing the fluorophore is not itself fluorescent and also, on subtraction from the measured spectra, eliminates the peaks that often occur from light scattering.

3.5. Spectrometer Parameters

When measuring fluorescence, the spectrometer has to be set up so that the correct information is recorded. In particular, the slit or bandwidths, experimental duration, and excitation and emission wavelengths have to be selected. All of these depend on the sample, desired signal to noise, and so forth. The slits are generally used to control the intensity, although some spectrometers are equipped with an iris aperture for this function. Care should be taken at low intensities to ensure that the excitation and emission slits are not open so wide that they overlap. The large slitwidths commonly used for fluorescence do not normally cause problems with resolution, since for most applications, spectra are broad single peaks with no fine structure. Excitation wavelengths are generally selected at or near the maximum absorption of the sample to ensure maximum fluorescence. In some cases, however, it may be required to excite samples some distance from the maximum to avoid other absorbance or fluorescence effects, e.g., the selective excitation of tryptophan over tyrosine in proteins (*see* Section 4.1.).

3.6. Anisotropy Measurements

When measuring anisotropy, the extent to which the optical system responds to polarized light must be taken into account. Monochromators in particular tend to transmit differing levels of light depending on its polarization. This effect must be corrected for to give accurate measurements. This is achieved by the *G* factor or the ratio of the efficiency of the optics to pass vertically compared to horizontally polarized light. Experimentally, the *G* factor is determined by:

$$G = (I_{HV}/I_{HH}) \quad (6)$$

where I_{HV} is the intensity of the fluorescence with the excitation polarizer in the horizontal position and the emission in the vertical position, and I_{HH} is the intensity with both polarizers in the horizontal position. The *G* factor can then be used to correct the anisotropy for the particular system in use by inserting it into the anisotropy equation as follows:

$$R = (I_{\|} - GI_{\perp}/I_{\|} + 2GI_{\perp}) \quad (7)$$

Note that the *G* factor applies for measuring both steady state and time-resolved anisotropies.

4. Applications of Fluorescence

The applications of fluorescence in biochemistry and molecular biology are wide and varied. Some of the more common are briefly mentioned here.

4.1. Intrinsic Protein Fluorescence

Proteins contain two amino acids with fluorescent side chains that, for practical purposes, are studied: tyrosine and tryptophan. The absorption and fluorescence spectra of tyrosine and tryptophan are shown in Fig. 3. Both tyrosine and tryptophan absorb maximally at 280 nm, although tryptophan has the higher absorbance at its maximum and shows significant absorbance at around 295 nm, whereas tyrosine does not. The fluorescence spectra measured at an exciting wavelength of 280 nm show that tyrosine and tryptophan have emission maxima at around 305 and 350 nm, respectively.

Tryptophan is by far the most important residue for fluorescence studies for two reasons: (1) Tryptophan side chain fluorescence is extremely sensitive to environment in general and in particular polarity. This means that tryptophans on the surface of proteins exposed to the polar solvent will have different emission maxima and intensity compared to the same residue buried deep inside a protein's strongly

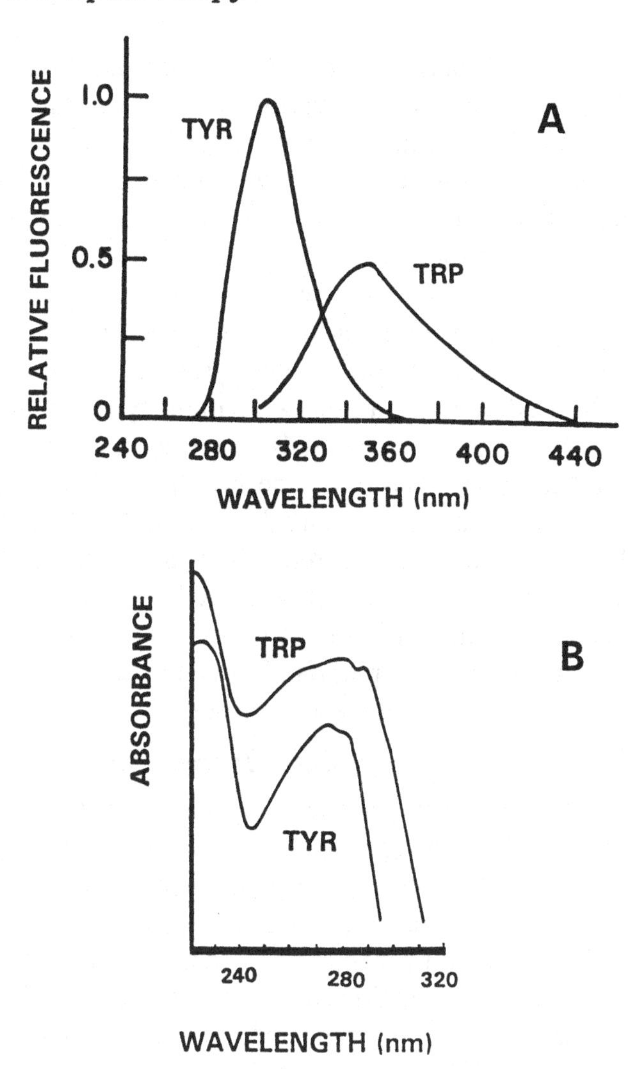

Fig. 3. Diagram showing the fluorescence emission spectra (**A**) and the absorption spectra (**B**) for tyrosine and tryptophan.

apolar core. For example, the emission maximum decreases in wavelength as the side chain is buried more deeply within the hydrophobic center of the protein. Tryptophan in solution (i.e., fully accessible to water molecules) has an emission maximum of 350 nm compared to the most buried (and therefore inaccessible to water) of tryptophan in the metalloprotein azurin, which has an emission maximum of <310 nm. (2) Tryptophan, where present with tyrosine in proteins, tends to dominate the spectrum, even though tyrosine is generally the more abundant residue. This is largely because energy transfer can occur between the excited tyrosine and the tryptophan, with the result being enhanced tryptophan fluorescence and decreased tyrosine fluorescence. Tyrosine is also easily quenched by groups common to amino acid side chains, further decreasing its fluorescence.

The result of the above effects is that tryptophan is much more widely studied as an intrinsic fluorophore than tyrosine (in fact, tyrosine is only generally studied when a protein contains tyrosine and no tryptophan). Proteins are generally studied by exciting at around the absorption maxima at 280 nm and observing the tryptophan-dominated spectra (if present). If, however, it is essential that only tryptophan fluorescence is observed, without interference from the tyrosine, the tryptophan in the protein may be selectively excited at 295 nm. This has the disadvantage of reducing the intensity of the fluorescence, but has the advantage of only exciting tryptophan (tyrosine does not absorb at that wavelength and therefore does not fluoresce).

4.1.1. Steady-State Fluorescence

The measurement of fluorescence of proteins can be used to study any phenomenon that causes changes to the environment of a tryptophan. This includes protein folding/unfolding, ligand or substrate binding, and association reactions. Generally, measurements of this type are made on a standard spectrofluorometer. Measurements of the emission spectra give information on the accessibility of the fluorophore to solvent characteristic of its environment. Changes in that environment will change both the emission maxima and the intensity of the fluorescence as the tryptophan is exposed/protected by solvent and quenched/unquenched by nearby groups. The intensity can be followed by measurement of the whole of the fluorescence (given by the area under the fluorescence spectra) or the fluorescence at a single wavelength. The change in emission maximum is followed by measuring the emission spectrum.

The steady-state anisotropy is often measured to follow the changes in the average correlation times of the protein under study. In this way, reactions that cause large size changes to the fluorescent protein, such as denaturation and association, can be followed.

4.1.2. Fluorescence Lifetimes

Like its steady-state spectra, the fluorescence lifetimes of tryptophan residues are also extremely sensitive to their environment, although owing to the equipment required and the complicated nature of the measurements, are rarely used to follow the basic phenomena that are listed above. The major application of fluorescence lifetimes to tryptophan in proteins is the study of protein dynamics. The fluorescence lifetimes of tryptophan in proteins varies between 1 and 6 ns, depending on its environment. Many of the functionally important structural fluctuations that occur in proteins occur on the same time scale. Fluorescence lifetimes are therefore a useful probe of these fluctuations.

Measurements of lifetimes can be used to measure energy-transfer distances and calculate real rate constants for fluorescence quenching (*see* Section 2.1.4.). The absolute rate of quenching provides information about the accessibility of the tryptophan to the quencher. If the quencher is neutral, such as acrylamide or oxygen, quenching of buried fluorophores can take place by the quencher "burrowing" into the interior of the protein. It is able to do this because of the fluctuating dynamic nature of the protein matrix. Measurement of the rate of quenching of buried tryptophans in proteins is therefore a useful measure of the fluctuations in the protein's structure *(6)*.

A lifetime spectrometer can also be used to study the rotation and changing environment of a tryptophan molecule using time-resolved anisotropy measurements. When the decay of anisotropy is measured on the nanosecond time scale, the decrease in anisotropy is owing either to the rotation of the whole protein or to the rotation of the tryptophan within the protein. Both of these often occur, which is reflected in the complex nature of the decay. Calculated correlation times from these components can therefore be used to access the size and shape of the protein, and also the rate of rotational freedom of a fluorophore within a protein, which would be expected to be dependent on the flexibility and packing of the protein's interior.

4.1.3. Problems Associated with Protein Fluorescence

The major problem with using tryptophan as a structural probe in proteins is the presence of more than one tryptophan, causing the fluorescence properties to be heterogeneous in nature. This complicates analysis and interpretation of the data, since changes measured in fluorescence properties will be of uncertain origin. Furthermore, fluorescence lifetimes of proteins with only a single lifetime are generally heterogeneous, i.e., the decays are multiexponential resulting from multiple lifetimes. This is thought to reflect the environment of the tryptophan, which could exist in many conformations with respect to its environment.

Much research has been carried out with the aim of resolving the fluorescence of individual tryptophan residues from others in the protein and investigating the origins of the heterogeneous lifetimes with some success. It should, however, always be remembered that the origins of fluorescence data are often more complicated than a single presented spectrum or lifetime suggests.

4.2. Applications of Extrinsic Fluorescence

Extrinsic fluorescence or fluorescence is generally the addition of fluorescent dyes to a system in order to study them. The use of such compounds generally reflects the limitations of the use of intrinsic fluorophores to study biological systems.

The high sensitivity of fluorescence is commonly utilized as a method of detection for small amounts of a substance, e.g., the detection of proteins/cells using fluorescently labeled antibodies for functions, such as ELISA, FACS, and microscopy. Such techniques rely on the ease of detection of small amounts of fluorophore to allow ease of quantitation in such systems.

The use of fluorescent compounds to study structure and dynamic systems is also extremely varied. Some examples of such structural and dynamic systems are briefly discussed below.

4.2.1. Extrinsic Probes of Protein Structure

Probes that are used to study protein structure and function can be either covalently or noncovalently bound to the protein. Of the covalently bound variety, such compounds as fluorescein and isothiocyanates are often bound to immunoglobulins for use in techniques, such as fluorescence microscopy. Dansyl derivatives, such as dansyl chloride and 8-{2-[(iodoacetyl)ethyl]amino} napthalenesulfonic acid (I-AEDANS), which bind to amino and sulfhydryl groups, respectively, are commonly used fluorescent probes in proteins. Such parameters as energy transfer between tryptophan and the dansyl group or decays of anisotropy (the dansyl group is much more suited to anisotropy measurements than tryptophan because of its longer lifetime) can be measured to give much more information concerning the structure and dynamics of the protein than would be possible without the label.

Probes that bind to proteins in a noncovalent manner can also be useful probes of structure. For example, 1-anilino-8-napthalenesulfonic acid (ANS) and related compounds are not fluorescent in an aqueous environment, but become highly fluorescent in a hydrophobic environment or when bound to a protein. ANS binds only to small hydrophobic patches on a protein's surface and, hence, generally does not bind unfolded or native proteins. It does however bind protein folding intermediates, some protein subunit interfaces and substrate binding sites. This therefore provides a useful method of following protein folding as well as some enzyme reactions and subunit associations.

4.2.2. Membrane Mobility

The incorporation of fluorescent dyes into lipid membranes and using anisotropy measurements to study their mobilities is a common method of studying their mobilities. Fluorescent dyes, which have long, thin hydrophobic structures that can be easily incorporated into the membrane, such as 1,6-diphenylhexatriene (DPH), are used.

These molecules, since they align themselves in the membrane, can rotate relatively rapidly along their long axis, which is inconsequential, since for most probes, the transition moment is not displaced. The molecules can also rotate along their short axis, but because of the length of the molecule and the constrictions of the membrane environment, this is very difficult. Such rotations will therefore become more frequent if the membrane becomes more flexible. This flexibility can be measured by measuring the anisotropy parameters of the molecule. Steady-state data are often measured as a function of viscosity (in a Perrin Plot) or as a function of temperature. The decay of anisotropy measured on the lifetime scale can be used to measure the rotation of such probes directly, often as a function of temperature. Probes are also available that contain different chemical groups to target specific regions of the bilayer relative to such parameters as chemical composition or membrane potential. This technique is therefore a powerful and relatively simple way of measuring the fluidity and flexibility of biological membranes.

4.2.3. Intracellular Ions

The study of intracellular ions, such as calcium, magnesium, and hydrogen, can be achieved by monitoring the fluorescence of certain fluorescent dyes. They change their fluorescence characteristics on binding of the relevant ion, allowing the concentration of that ion to be accurately followed throughout an experiment. For example, when FURA-2, a polycyclic calcium chelator, binds calcium ions, the excitation maximum shifts from 380 to 340 nm, allowing the concentration of calcium ions to be determined in the nanomolar to micromolar concentration range.

5. Summary

This chapter has introduced the theory of fluorescence, and briefly shown that it is a technique with many wide and varied applications in biochemistry. It has the overall advantage of being able to carry out measurements on small amounts of sample in solution, often revealing important information concerning the structure and dynamics of these systems.

References

1. Lakowicz, J. R. (1983) *Principles of Fluorescence Spectroscopy.* Plenum, New York. Detailed, comprehensive work covering theory and practice.
2. Eftink, M. R. and Ghiron, C. A. (1980) Fluorescence quenching studies with proteins. *Anal. Biochem.* **114,** 199–227.
3. Stryer, L. (1978) Energy transfer of fluorescence as a molecular ruler. *Ann. Rev. Biochem.* **47,** 819–846.
4. Brand, L. and Withold, B. (1967) Fluorescence measurements. *Methods Enzymol.* **11,** 776–856. Old but still relevant introduction to theory and practice.
5. Harris, D. A. and Bashford, C. L. (eds.) (1987) *Spectrophotometry and Spectrofluorometry a Practical Approach.* IRL, Oxford, UK. Volume concentrating on the practical aspects of fluorescence.
6. Demchenko, A. P. (1988) Fluorescence measurement of protein dynamics. *Essays Biochem.* **22,** 120–157.

CHAPTER 16

Circular Dichroism

Alex F. Drake

1. The Technique and Instrumentation

Circular dichroism (CD) is the difference in the absorption of left and right circularly polarized light. It can be considered as the absorption spectrum measured with left circularly polarized light minus the absorption spectrum measured with right circularly polarized light. From Beer's law:

$$\Delta\varepsilon = \varepsilon_L - \varepsilon_R = (A_L - A_R)/cl = \Delta A/cl \qquad (1)$$

where $\Delta\varepsilon$ is the differential molar extinction coefficient; ΔA is the differential absorbance between left circularly polarized (A_L) and right circularly polarized light (A_R); c is the concentration in moles per liter; and l is the cuvet pathlength in centimeters. The differential is minute ($\Delta A = 10^{-3}$–10^{-5}) and cannot be measured simply, by difference, with a modified ordinary UV/Vis spectrophotometer. A specialized instrument is required known as a CD spectrometer, spectropolarimeter, or dichrograph.

1.1. The Instrument

As illustrated in Fig. 1, light from an intense source, rendered both monochromatic and linearly polarized by the monochromator, passes through a polarization modulator to the photomultiplier detector. The polarization modulator induces a periodic variation in the polarization of the light beam through all ellipticities from left circular through

From: *Methods in Molecular Biology, Vol. 22: Microscopy, Optical Spectroscopy, and Macroscopic Techniques* Edited by: C. Jones, B. Mulloy, and A. H. Thomas

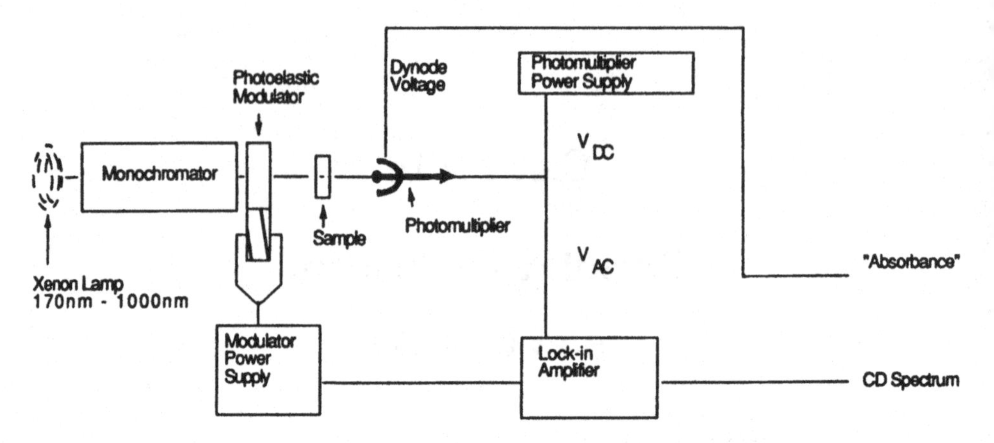

Fig. 1. The CD spectrometer.

elliptical, unchanged linear, and elliptical to right circular. During this cycle (at any fixed wavelength) the intensity of the light beam does not vary. The introduction of an optically active sample, which absorbs at this wavelength, sees a preferential absorption during one of the polarization periods, and the intensity of the transmitted light will now, therefore, vary during the modulation cycle. This intensity variation is directly related to the circular dichroism of the sample at the specified wavelength. Successive detection at various wavelengths leads to the generation of the full CD spectrum.

The signal from the photomultiplier detector is represented in Fig. 2 with the ordinary absorbance given as:

$$A = \log(I_o/I_t) = \log(V_o/V_{DC}) \quad (2)$$

The circular dichroism is given, assuming a fixed I_o, as:

$$\Delta A = (A_L - A_R) = \log(I_o/I_L) - \log(I_o/I_R) = \log(I_R) - \log(I_L) \quad (3)$$

In practice over all wavelengths, I_o varies owing to lamp output and monochromator characteristics; in regions of absorption the transmitted light intensity is dependent on sample absorption. Legrande and Grosjean *(1)*, in describing their patented design of the modern CD spectrometer involving polarization modulation *(2)*, showed that for very low values of $(I_R - I_L)$, CD is given by:

$$\Delta A = (A_L - A_R) = (I_R - I_L)/(I_R + I_L) = V_{AC}/V_{DC} \quad (4)$$

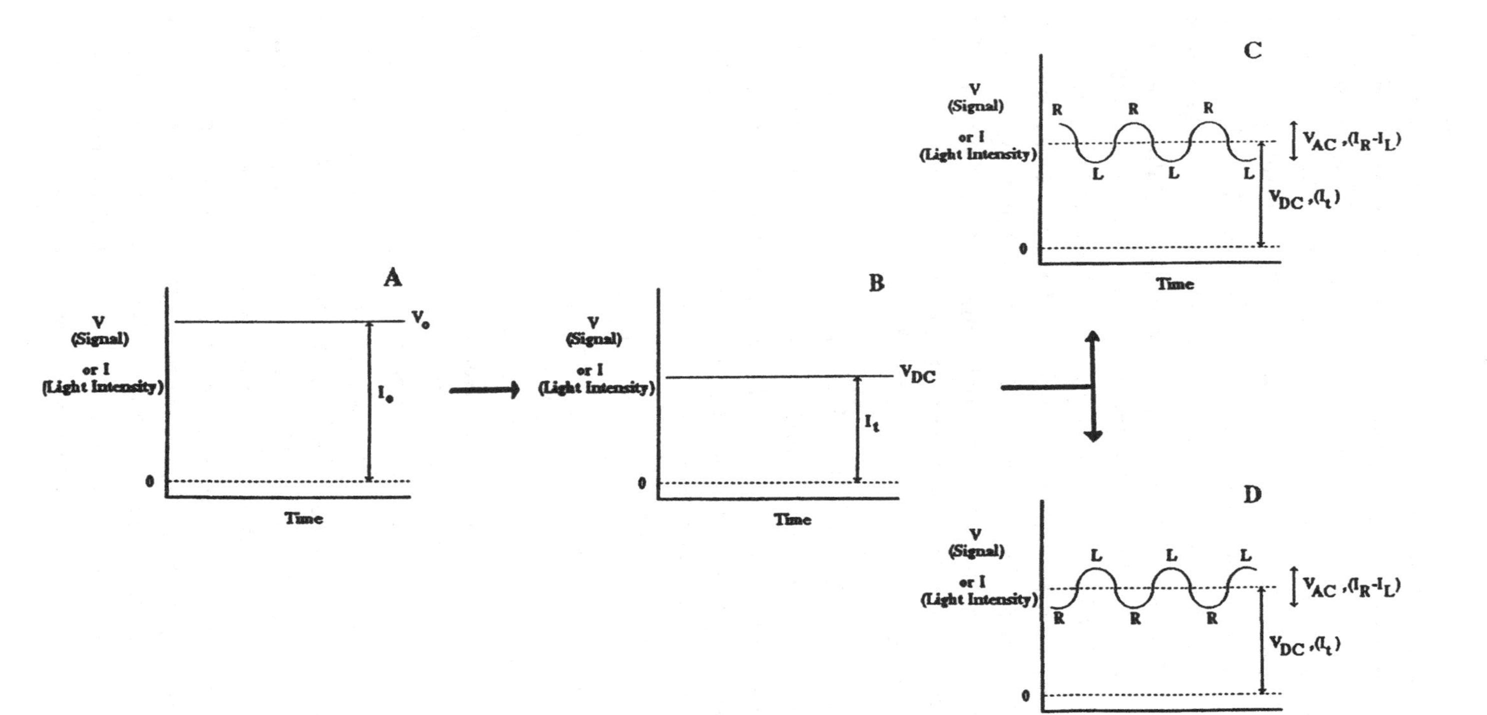

Fig. 2. The CD spectrometer signal. **(A)** the signal coming in from the photomultiplier at a fixed wavelength in the absence of sample, **(B)** in the presence of an absorbing optically inactive sample, **(C)** in the presence of an absorbing sample with positive circular dichroism, and **(D)** in the presence of an absorbing sample with negative circular dichroism.

Again V_{AC} is too small for the V_{AC}/V_{DC} ratio to be easily measured directly, rather it is measured as the V_{AC} associated with a constant V_{DC}. The constant V_{DC} is maintained by a feedback servo system that varies the high voltage (dynode voltage) on the photomultiplier so that both V_{AC} and V_{DC} are amplified to the same extent such that V_{DC} is constant and V_{AC} is proportional to ΔA over all measurement wavelengths, conditions, and concentrations. $\Delta A = (A_L - A_R)$ is determined as the rectified V_{AC} signal from the detector with the electronics of the instrument phase-locked to the polarization modulation via the power supply of the modulator. The resulting varying dynode voltage is a good measure of the light level (V_{DC}), and being approximately proportional to log(I) can provide a measure of the ordinary absorbance, A. This signal is often registered as a supplementary channel on the CD spectrometer providing a useful indication of the overall light throughput, and hence is a good guide of instrument condition and sample suitability.

1.2. Measuring CD Spectra

Several parameters need to be controlled in a CD measurement: the concentration and pathlength of the sample and the instrumental settings of spectral bandwidth (spectrometer slitwidth), time constant, scan speed and digital data resolution.

1.2.1. The Sample

The higher the light level striking the detector, as registered by a low dynode voltage on the photomultiplier, the lower the noise. At the same time, the more sample there is (higher concentration and longer pathlength), the greater is the CD signal. More sample (concentration or pathlength) means greater absorption, lower light throughput, and thus more noise. Theoretically, an absorbance of $A = 0.864$ is seen to provide an optimal balance giving the best signal-to-noise ratio. This applies to the total absorbance of sample and solvent. Therefore, the recommended conditions are a cell pathlength that ensures a low solvent absorption (≤0.5 mm to reach 185 nm) with a sample concentration in this cell such that the "cell + sample + solvent" absorption never exceeds ~1.4 at wavelengths >200 nm. A total maximum absorbance of ~1.0 is the target. For far UV measurements, 260–185 nm, the total absorbance of "cell + sample + solvent" should never exceed 0.8. These absorbance criteria are also important to ensure that stray light does not lead to severe signal distortion as it can, particularly near the wavelength limit of the spectrometer (185 nm). These imposed measurement conditions should ideally be controlled by separate measurements on an ordinary UV/Vis spectrophotometer.

1.2.2. Instrument Settings

1.2.2.1. Spectral Bandwidth (Instrument Slitwidth)

Spectral bandwidth is a measure of the purity of the wavelength falling on the sample. Commercial instruments now operate so that during the measurement, the monochromator slitwidth varies continually to maintain a constant spectral bandwidth. A value of 1 nm is conventional ensuring that at all wavelengths the "spread" of light is ±0.5 nm of the given value. This is good enough for most measurements. Fine structure, such as that associated with phenylalanine, may benefit from a smaller spectral bandwidth (0.5 nm). However, as halving spectral bandwidth reduces light level by a factor of four (entrance and exit slits halved), a substantial increase in noise is the result that may defeat the object of improving the spectrum. At the limits of the spectrometer where performance is falling off, increasing the spectral bandwidth offers the advantage of lower noise. A spectral bandwidth of 2–4 nm is quite acceptable at 600 nm and beyond. In the far UV (260–185 nm) a spectral bandwidth of 2 nm can be used to good effect to reduce noise, avoiding the need for signal accumulation (*see* Section 1.2.2.2.), although strictly speaking for good secondary structure analysis a 1 nm spectral bandwidth is to be preferred.

1.2.2.2. Time Constant and Scan Speed

In the days of analog measurements, time constant was the damping factor imposed on the signal. This took the form of a resistance/capacitance (RC) filter immediately prior to data presentation typically on a chart recorder. The choice of R and C, selected by a switch on the instrument console, effectively controlled the time required for a signal change to arrive at its true value. Thus a 4-s time constant meant that it took 4 s for the recorder to register 0.632 of the target value. Under this dampening, fast changes caused by noise are suppressed. The spectrometer scan speed must be slow enough to allow time for the measured signal to be recorded faithfully. Higher time constants give greater noise suppression but will require longer scan times. Typically a scan speed of 10 nm/min is required for a faithful recording with a 4-s time constant. Noise reduction is proportional to the square root of the time constant; therefore noise is successively halved by changes in time constant from 1 to 4 s and 4 to 16 s. In normal practice, time constants faster than 1 s (with associated higher noise) are only used to monitor fast processes, time constants greater than 16 s (e.g., 64 s) require very slow spectrum measurements which are better achieved by spectrum accumulation.

Modern instruments are based on digital signal processing. Thus, rather than time constant, the term *integration time*, derived from digital filtering, is used. Again this must be considered when choosing a scan speed.

An important means of reducing noise is the taking of the average of several repeated measurements. This requires computer data handling. Thus the noise associated with a single measurement can be halved by taking the average of four measurements. A fourfold noise reduction requires 16 scans. The accumulation of 1, 4, 16, or 64 scans are the logical options.

As spectra are stored digitally on a computer, it is necessary to decide on the digital resolution. Registering data every 0.2 nm is reasonable, certainly for typical spectra in the range 260–185 nm, giving 376 data points. This leads to the newer concepts of spectral measurement. How long should the instrument remain accumulating data at each of these wavelengths? Thus a variable speed can be employed with the spectrometer spending a longer time at those wavelengths where the data is noisy owing to low light levels.

Postmeasurement mathematical smoothing or Fourier transform filtering are essentially cosmetic, can induce distortions, and should be applied strictly for presentation purposes. If the sample has been correctly presented to a correctly set spectrometer, the modern instruments will normally ensure good measurements in a single scan, although there are occasions where signal averaging may be advantageous. Subsequent smoothing is a last resort.

1.3. CD Units

Prior to the patented technique of Legrande and Grosjean *(1)* discussed earlier, the routine measurement of the small ΔAs required for practical use was not possible. Accordingly, CD was supplanted as the technique of choice by optical rotation measurements that were feasible. At this time, the CD measurements that were made were based on a consequence of the differential absorption of left and right circularly polarized light leading to the change in ellipticity of incident linearly polarized light as it passed through an absorbing, optically active sample. These measurements were achieved by the rotation of an optical element in the optical train of the instrument to achieve an optical null. This mechanical rotation, measured in degrees, related to the ellipticity change and was quoted as the CD. Molar ellipticity is given as:

$$[\Theta] = \theta/cl \tag{5}$$

where θ is the measured ellipticity, c is the concentration in decimoles per liter, and l is the pathlength in centimeters. The units are degree $\cdot$ cm^2 $\cdot$ dmol^{-1}. This is a redundant unit and its use is to be strongly discouraged. The true measure of CD defined earlier has the units M^{-1}cm^{-1}. The relationships between the two units are:

$$[\Theta] = 3300\,\Delta\varepsilon \tag{6}$$

$$\theta = 33{,}000\,\Delta A \tag{7}$$

2. Applications

To exhibit CD a sample must be optically active, which in turn requires that the molecule is not superposable on its mirror image. The existence of such chiral molecules is critical to the chemical aspects of living systems. For a right-handed α-helix the constituent amino acids must be L– (more correctly S–), for nucleic acids to be right-handed the ribose units need to be D–. It is this monomer homochirality in nature that allows the generation of the macrostructures that lead to replication, and the determination of this stereochemistry is extremely important. X-ray crystallography offers the only means of achieving this with complete security. However, crystals of the chemical compound are needed, which is often either difficult or not convenient. CD is the spectroscopic technique that is a direct consequence of the absolute spatial aspect of molecular shape *(3,4)*. In principle, the sign (and magnitude) of a CD band associated with or, more correctly, deriving from a particular transition needs to be correlated with structure. This involves theoretical calculations or more simply the comparison of the CD of the com-

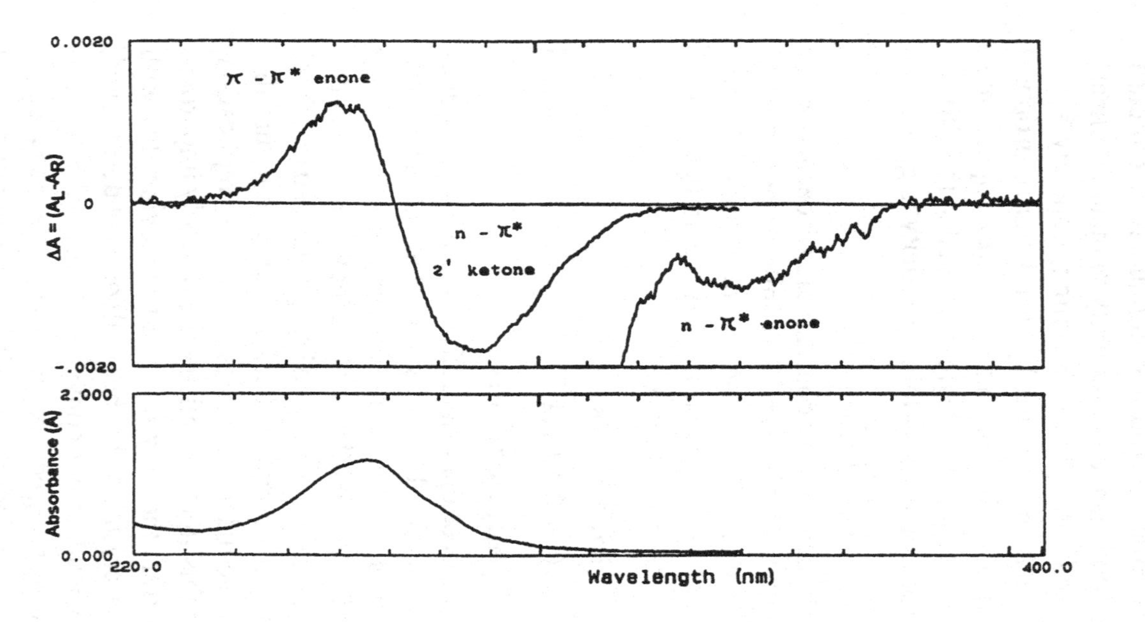

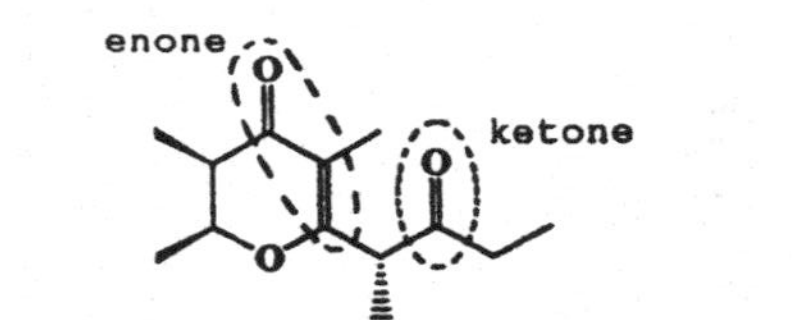

Fig. 3. The CD spectra of natural stegobinone, a beetle pheremone, with (2R, 3R, 1'R) stereochemistry. The negative CD around 289 nm reflects the local chirality around the aliphatic ketone, whereas the 340 and 261 nm bands reflect the enone environment. The ordinary UV maximum is at 267 nm.

pound under study with that of a well chosen compound of previously established stereochemistry. An example is illustrated in Fig. 3.

The remainder of this chapter concentrates on the CD associated with the conformation (secondary structure) of biological macromolecules, discussed with respect to the optical activity imposed by the optically active monomer units. Two types of CD can be distinguished. First, there is CD related to the biopolymer backbone (amides in proteins and peptides; the bases in nucleic acids), which is derived from amide–amide or base–base interactions. Superposed on this is the optical activity of prosthetic groups (chromophores), such as the phenol of tyrosine, the phenyl of phenylalanine, the indole of tryptophan, the disulfide bond, bound ligands, heme groups, and the like, which sense the chirality of the macromolecule and hence exhibit CD or have their own CD modified if they are optically active themselves. The CD will be associated with the absorption spectrum of the chromophore in question, will be characteristic of its location and orientation with respect to the macromolecule, and may overlay the backbone spectrum.

2.1. Protein Secondary Structure

The basic chromophore of the polypeptide backbone is the amide group that has two electronic absorptions, an electric dipole allowed (magnetic dipole forbidden) $\pi - \pi^*$ transition giving a strong ordinary absorption ($\varepsilon \sim 10{,}000$) around 190 nm, and a magnetic dipole allowed (electric dipole forbidden) $n - \pi^*$ transition giving a weak ordinary absorption ($\varepsilon \sim 100$) around 210 nm, which is often masked by the $\pi - \pi^*$. This defines the spectral range, 250–170 nm, for secondary structure analysis. The $\pi - \pi^*$ and the $n - \pi^*$ transitions become optically active under the influence of the substituents on the asymmetric α-carbon atom in a free amino acid amide (*see* Fig. 4). In the polypeptide, the amide groups interact with each other to provide a CD spectrum that is more characteristic of the amide–amide orientation than the monomer stereochemistry. This dominates the CD contributions induced by the centers of chirality.

The optical activity derived from coupling chromophores is related to the relative orientations of the transition moments and hence the secondary structure (conformation) of the polypeptide chain (Fig. 4). An example of how this applies to the α-helix is presented in Fig. 5.

THE MONOMER UNIT: Optical activity induced by the groups attached to the centre of chirality ($n - \pi^*$ is about 0 atom, it is offset for clarity)

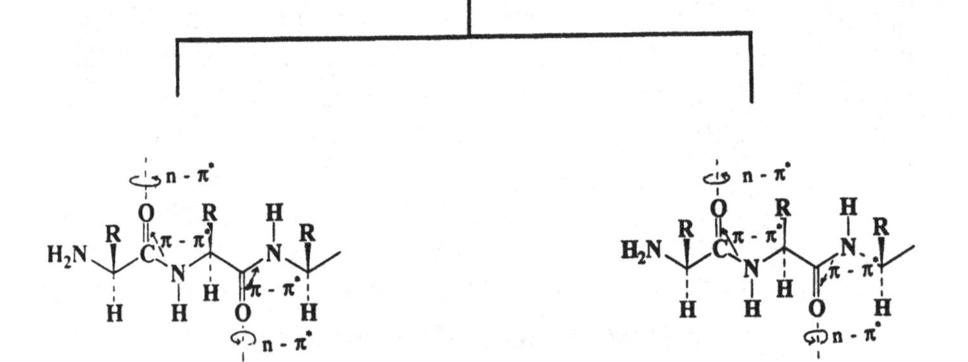

High energy coupling mode

Low energy coupling mode

THE POLYPEPTIDE:
Optical activity induced by the coupling of adjacent $\pi - \pi^*$ transitions (two coupling modes are possible)

Fig. 4. Protein and peptide optical activity.

Despite many attempts, a reliable calculation of the CD associated with a specific protein (peptide) structure from first principles remains difficult. Suffice it to appreciate that different conformations have different amide–amide orientations and hence different CD spectra (Fig. 6). In practice, reference to models or correlation of the CD spectra of proteins with known X-ray structures has led to a consensus set of spectra, which can be treated as a series of fingerprints (*see* Fig. 7, *see* refs. *5–7*).

Therefore, confronted with the CD of a typical protein (Fig. 8, *see* ref. *8*), one can attempt to reduce the measured spectrum to a linear combination of fundamental spectra. At first, Greenfield and Fasman

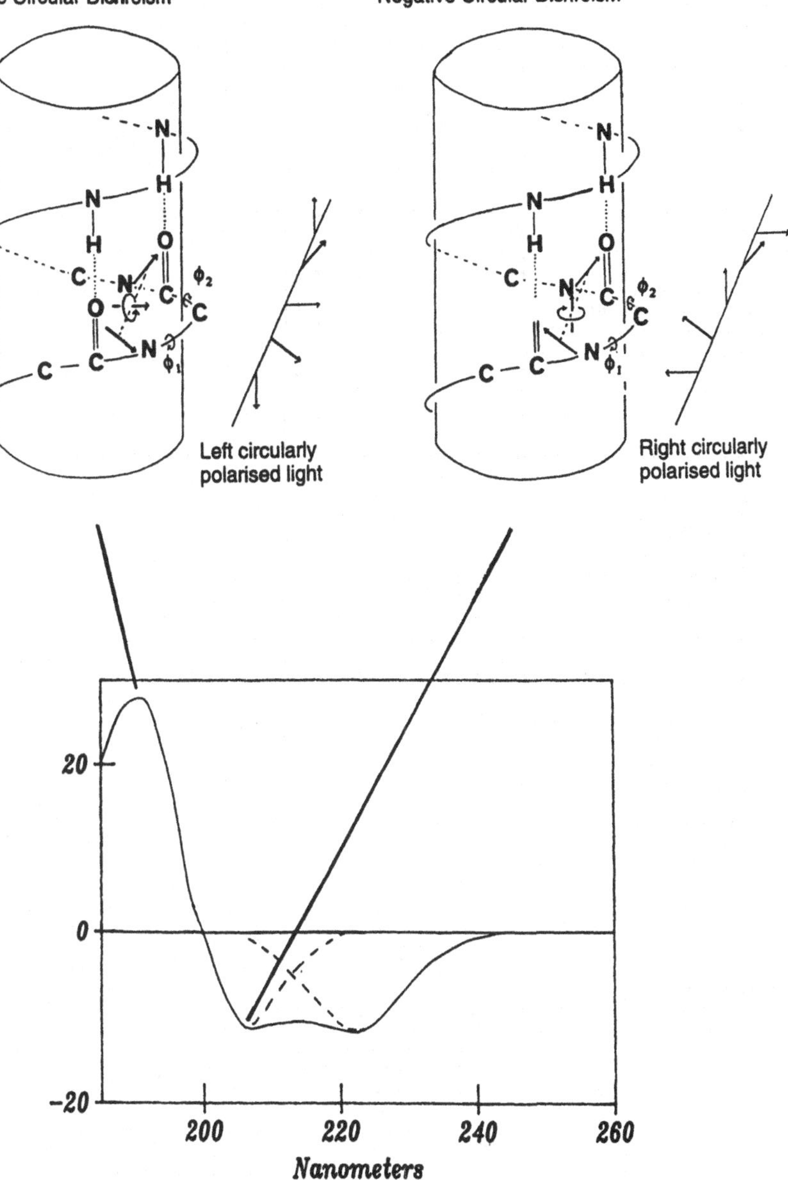

Fig. 5. The electronic origin of circular dichroism in the α-helix.

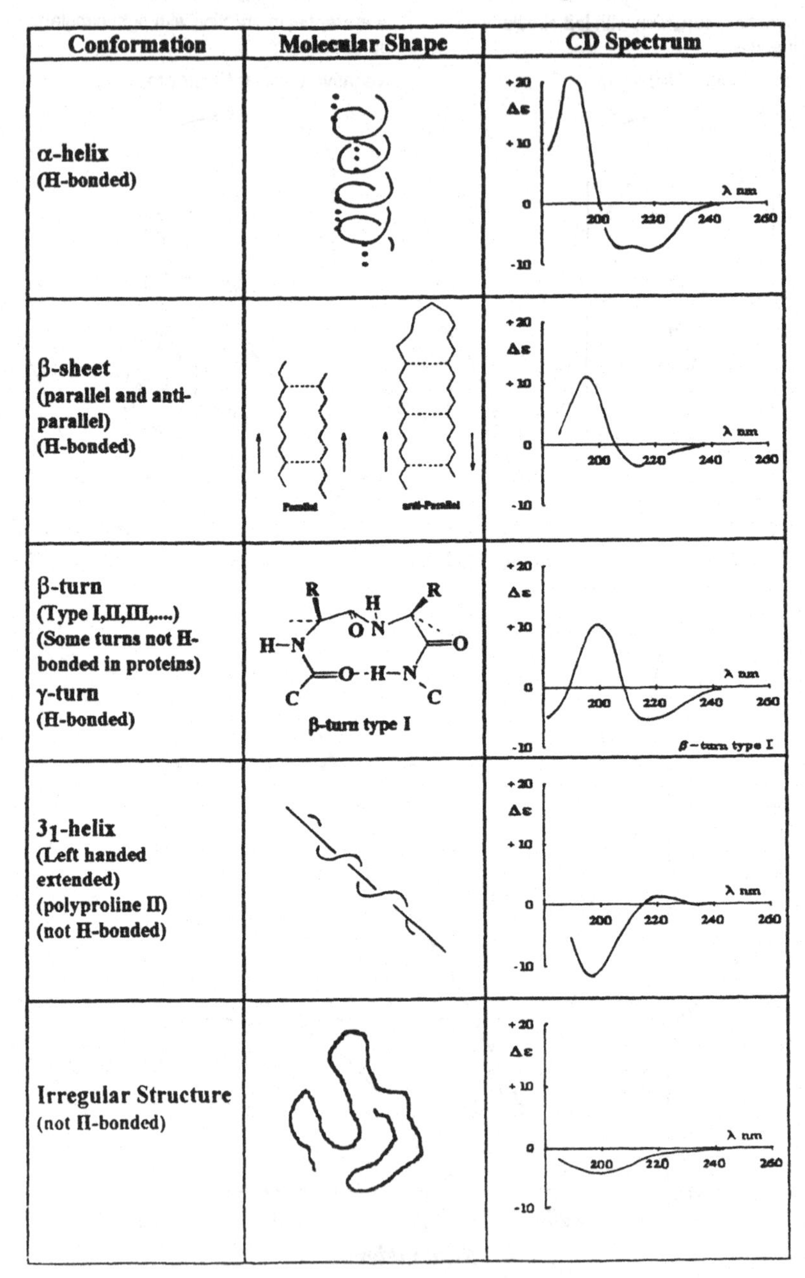

Fig. 6. The major secondary structure classes and their associated CD spectra.

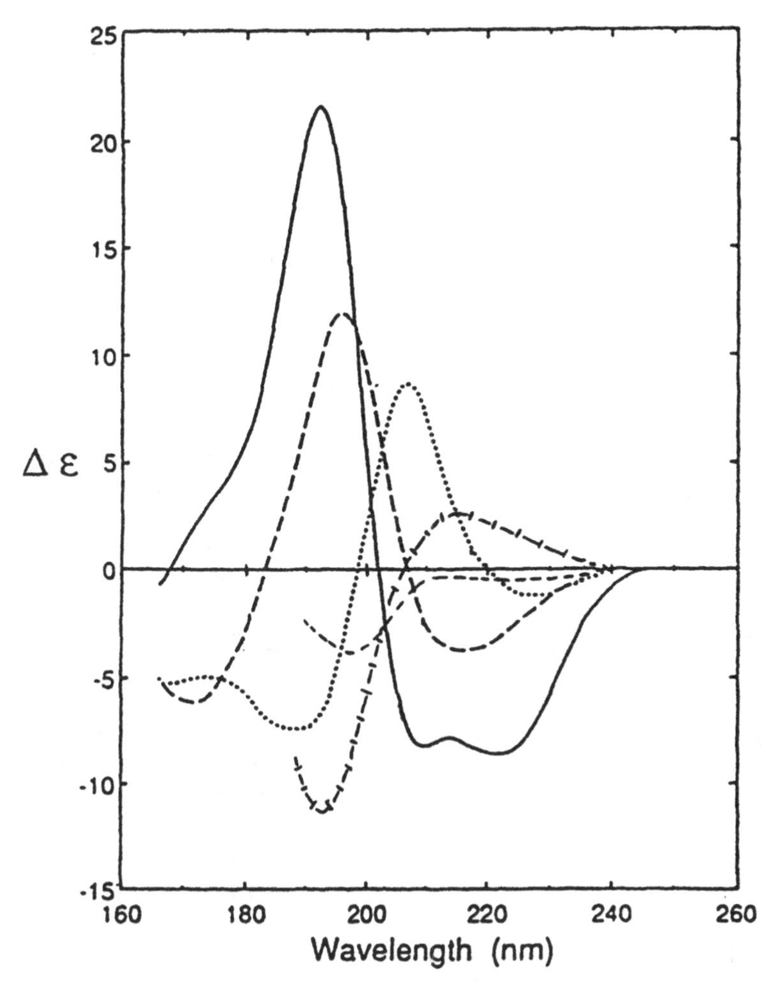

Fig. 7. CD spectra associated with various secondary structures: α-helix (——), antiparallel β-sheet (----) turn type 1 (······), and left-handed extended 3_1-helix (–I–I–I–). Redrawn using data from refs. *5–7*.

(9) considered the CD of a protein ($\Delta A^{\lambda}{}_{obs}$) observed at every wavelength was given by the expression:

$$\Delta A^{\lambda}{}_{obs} = \Delta A^{\lambda}{}_{\alpha\text{-helix}} + \Delta A^{\lambda}{}_{\beta\text{-sheet}} + \Delta A^{\lambda}{}_{\text{random}} \qquad (8)$$

where ($\Delta A^{\lambda}{}_{\text{element}}$) is the CD contribution of the indicated structural component.

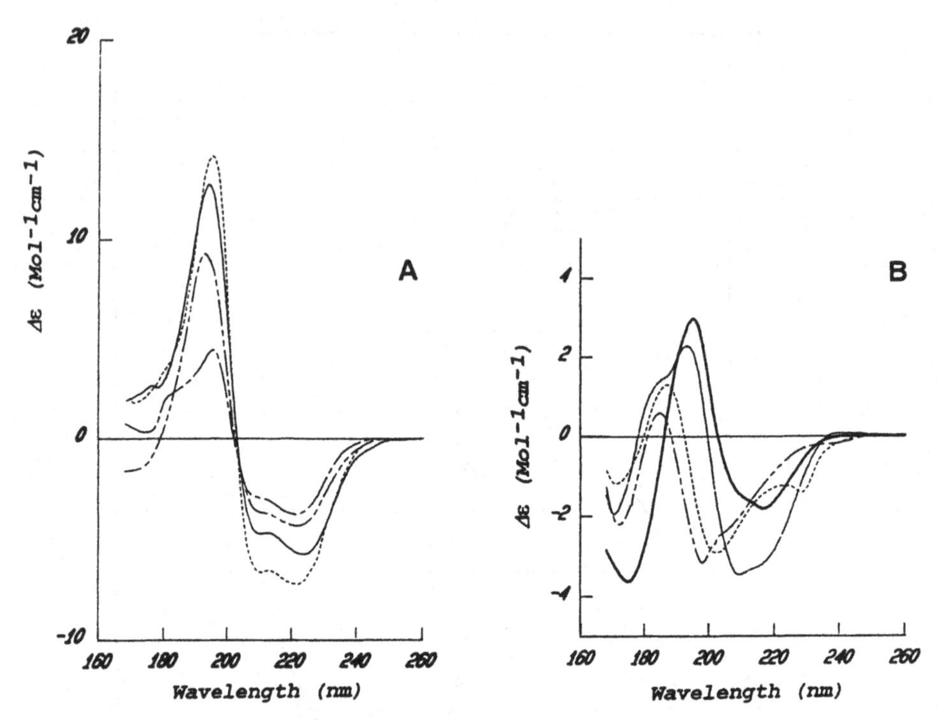

Fig. 8. Representative CD spectra of proteins with secondary structure content defined by X-ray crystallography (reproduced from ref. *8*). **(A)** α-helix rich: (——) T4 lysozyme (67%H, 10%S, 6%T, 17%O), (----) hemoglobin (75%H, 0%S, 14%T, 11%O), (— - —) cytochrome c (38%H, 0%S, 17%T, 45%O), (— -- —) lactate dehydrogenase (chicken heart) (41%H, 17%S, 11%T, 31%O). **(B)** β-sheet rich: (——) prealbumin (97%H, 45%S, 14%T, 34%O), (----) α-chymotrypsin (10%H, 34%S, 20%T, 36%O), (— -- —) elastase (10%H, 37%S, 22%T, 31%O), (——) ribonuclease A(24%H, 33%S, 14%T, 29%O). Conformation contents are given by H = α-helix, S = β-sheet, T = β-turn, and O = other.

Applying Beer's Law to each element, $\Delta A = (\varepsilon^{\lambda}_{element}) \cdot (c_{element}) \cdot l$, where $c_{element}$ is the concentration of the indicated secondary structure component and l is the measurement cell pathlength, gives:

$$\Delta A^{\lambda}_{obs}/l = \Delta\varepsilon^{\lambda}_{\alpha\text{-helix}} \cdot c_{\alpha\text{-helix}} + \Delta\varepsilon^{\lambda}_{\beta\text{-sheet}} \cdot c_{\beta\text{-sheet}} + \Delta\varepsilon^{\lambda}_{random} \cdot c_{random} \quad (9)$$

with a total protein concentration, $c_{protein}$, the percentage of each secondary structure component is given by the respective $(c_{element}/c_{protein}) \cdot$ 100% having taken the Δεs from the reference data set. To solve this equation, data from at least three wavelengths is needed. In practice, ΔA values were taken at several wavelengths and fed into a computer program that effectively solved the set of simultaneous equations to extract the best $(c_{element}/c_{protein}) \cdot 100\%$ values. However, it soon became apparent that this approach was too simplistic, and explicit contributions from other structural elements ought to be included. Various authors have indicated that:

1. Two general classes of β-sheet, parallel and antiparallel, need to be considered.
2. There are various β-turns with different CD profiles, although some writers erroneously group them together as a single component.
3. The designated CD spectrum of the "random coil" has recently been reinterpreted as a mixture of contributions from the polyproline II type, left-handed helix conformation, and what is now better termed the unordered or irregular conformation.

Equation 8 now becomes:

$$\begin{aligned}\Delta A^{\lambda}_{obs} = {} & \Delta A^{\lambda}_{\alpha\text{-helix}} + [\Delta A^{\lambda}_{\text{parallel } \beta\text{-sheet}} + \Delta A^{\lambda}_{\text{antiparallel } \beta\text{-sheet}}] \\ & + [\Delta A^{\lambda}_{\beta\text{-turnI(III)}} + \Delta A^{\lambda}_{\beta\text{-turnII}} + \Delta A^{\lambda}_{\gamma\text{-turn}} + \ldots \\ & + [\Delta A^{\lambda}_{LH} + \Delta A^{\lambda}_{\text{irregular}}] \end{aligned} \tag{10}$$

Thus the complete solution requires the extraction of information relating to up to eight conformational components. A daunting task indeed from, at most, 80 nm (250–170 nm) of a single UV CD spectrum. This reinforces the need for pure proteins dissolved in transmitting solvents, as an impure protein extract dissolved in phosphate-buffer-saline may in fact only transmit as far as 200 nm, supplying only 50 nm of data. This analysis also presupposes that there exist unique spectra for the structure types independent of length and distortion, and that each structure type is well defined. It is not surprising, therefore, that there have been several attempts to provide a computer package that successfully and confidently analyzes a CD spectrum in terms of secondary structure component contributions. Fortunately, most proteins have a dominant conformational feature, α-helix or β-sheet, allowing a reasonable estimate of this single component, to a precision of 5% for higher α-helix contents in good cases. Some computer programs restrict output to an estimation of α-helix, β-sheet, and remainder only. These concepts have been reviewed by Yang et al. *(10)* with the most recent work being that of W. C. Johnson *(8,11)*.

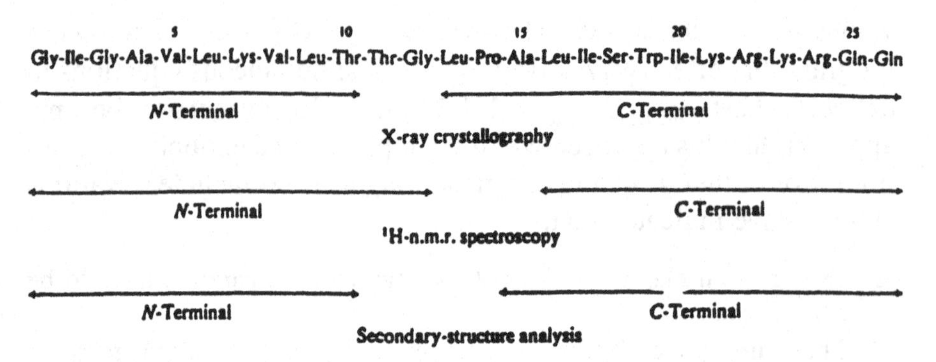

Fig. 9. The amino acid sequence of melittin. The extents of the N- and C-terminal helices as determined by X-ray crystallography *(13)*, NMR spectroscopy *(12)*, and structural prediction *(14)* with use of Levitt *(15)* parameters are shown.

Rather than attempt an accurate calculation, the CD spectra should be used to: (1) set a protein structure within a class of structures, e.g., α-helix rich, β-sheet rich, α + β, or α/β proteins; and (2) monitor processes such as protein unfolding in order to better define them providing a basis for other, perhaps more refined, studies such as those involving NMR spectroscopy.

Without question, CD does provide an excellent method of following changes in secondary structure, even if the changing states are not themselves completely defined. Figures 9–11 show the results of a CD study on the bee venom melittin. The sequence and secondary structure, as determined by NMR spectroscopy *(12)*, X-ray crystallography *(13)*, and prediction algorithms are shown in Fig. 9. In pure water at pH 7, CD shows that melittin has little ordered structure (Fig. 10), but addition of salts cause a dramatic induction of α-helix. The effect is anion specific, and addition of different sodium salts induces the α-helix at different concentrations. The use of phosphate as a buffering agent can induce structural changes relative to, e.g., Tris (which has chloride as a counter ion). The α-helix is also induced by interaction with sodium dodecylsulfate or egg lecithin vesicles (Fig. 11).

The CD of α-cobratoxin (Fig. 12) in the near UV (310–240 nm) derives from transitions localized in the prosthetic groups. The variation of CD in this region can be used to monitor the changes in its the conformation and local environment owing to variations in conditions, such

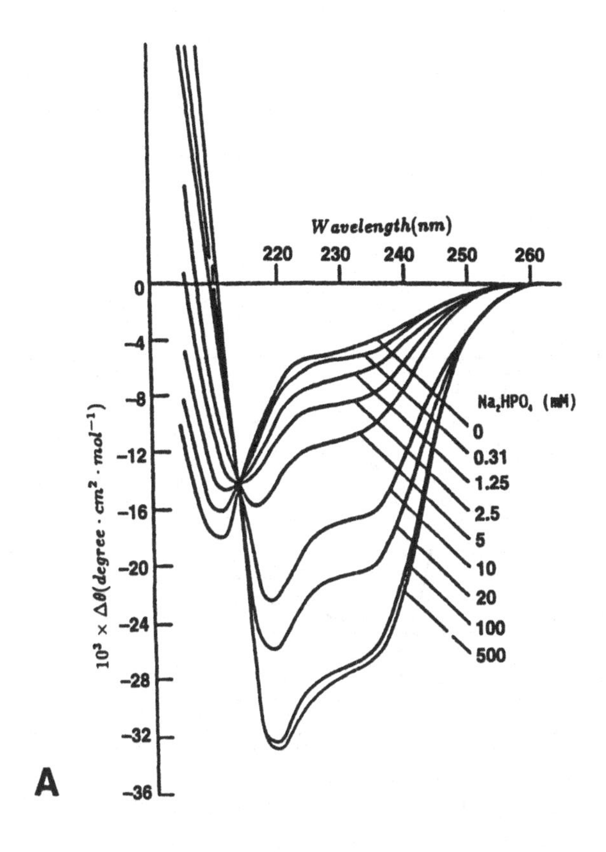

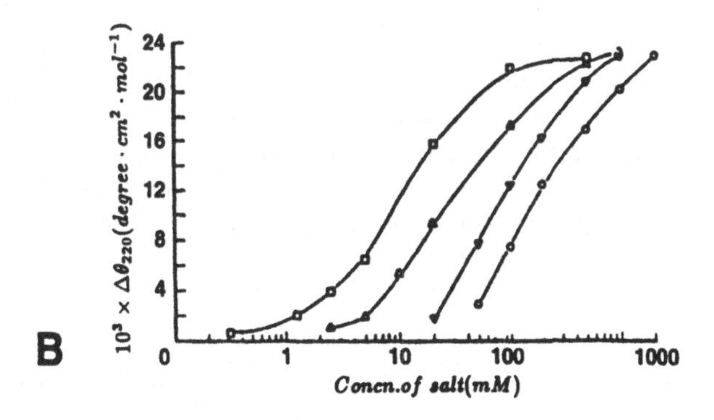

Fig. 10. **(A)** Influence of phosphate concentration on the CD spectrum of melittin at pH 7.4. The concentration of melittin was 0.5 mg/mL. **(B)** CD ellipticity of melittin (220 nm) as a function of salt concentration at pH 8.0. For details, *see text.* (□) Na_2HPO_4; (△) Na_2SO_4; (▽) $NaClO_4$; (○) NaCl.

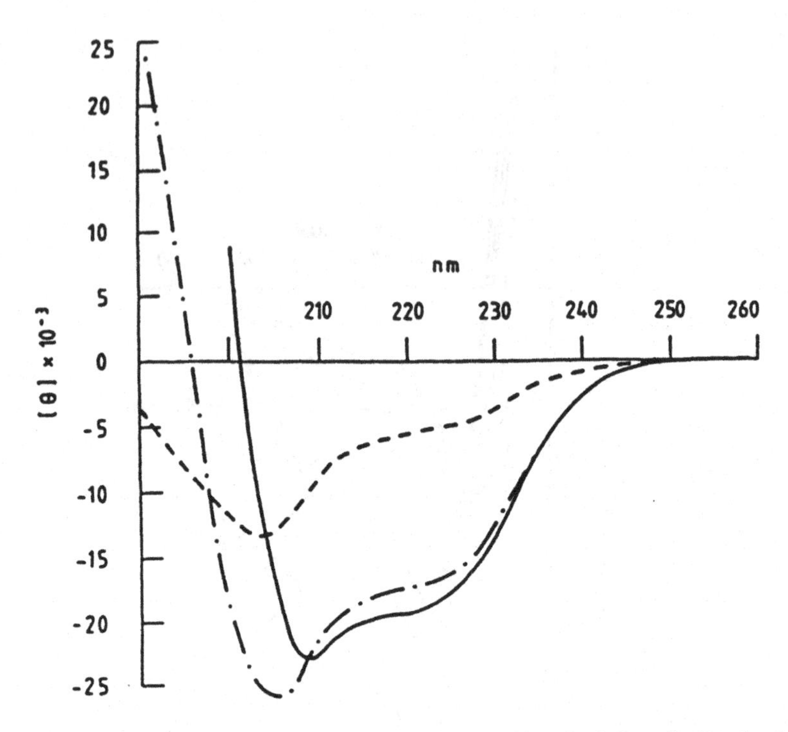

Fig. 11. The CD of melittin (0.5 mg/mL, Sigma Chemical Co., St. Louis, MO) dissolved in sodium phosphate (0.15*M*, pH 7.4) ——; dissolved in Tris-HCl (10 m*M*, pH 7.4) ---- and melittin (0.4 mg/mL) dissolved in Tris-HCl (10 mg/mL, pH 7.4) containing liposomes prepared from egg yolk phosphatidylcholine (1.3 mg/mL) —·—·—. The liposome preparation was sonicated for 5 min under N_2 and centrifuged at 10,000*g* for 10 min prior to the addition of melittin. This centrifugation step ensured that only small vesicles were present in the preparation, so distortion caused by scattering over the 205–260 nm range was minimal *(17,18)*. In the absence of melittin, the baseline was not significantly changed by the presence of liposomes in the 0.1 mm pathlength cell. The number of membrane shells, per liposome, was in the range 5–10. The melittin:phospholipid ratio was 1:13. The spectra were obtained on a Jasco J40CS using 0.1 and 1.0 mm cells at 20°C. At this temperature, melittin did not cause liposome fusion. The results are expressed in terms of molar ellipticity based on an average monomer mol wt of 110; the units are degree · cm^2 · mol^{-1}.

as pH, temperature, and the like. In Fig. 13, the variation in temperature causes a change in the tryptophan and its local environment (seen in the near UV spectrum) well before the onset of backbone denaturation, observed in the far UV. Figure 14 shows the variation of the far UV CD spectrum caused by changes in pH. Structural changes here result from

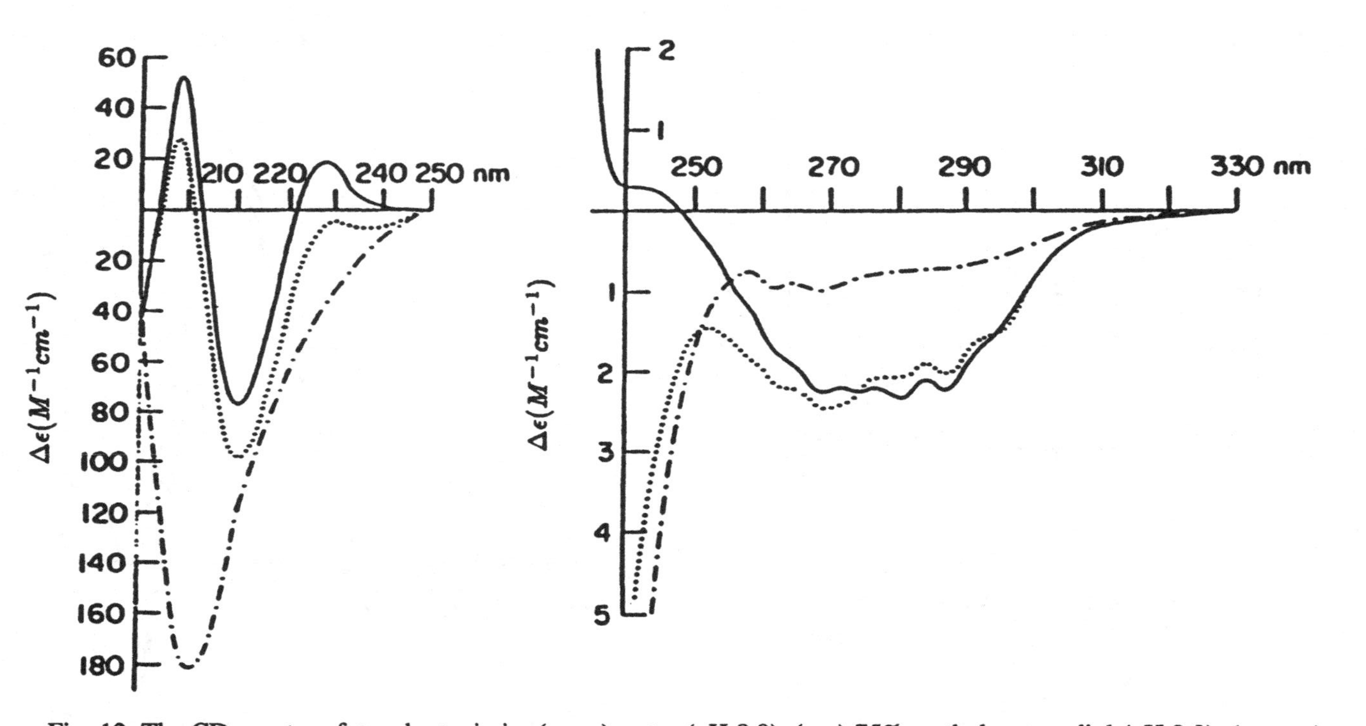

Fig. 12. The CD spectra of α-cobratoxin in: (——) water (pH 8.0); (·····) 75% methylpentanediol (pH 2.0); (–·–·–) sodium dodecyl sulfate (10 mg/mL).

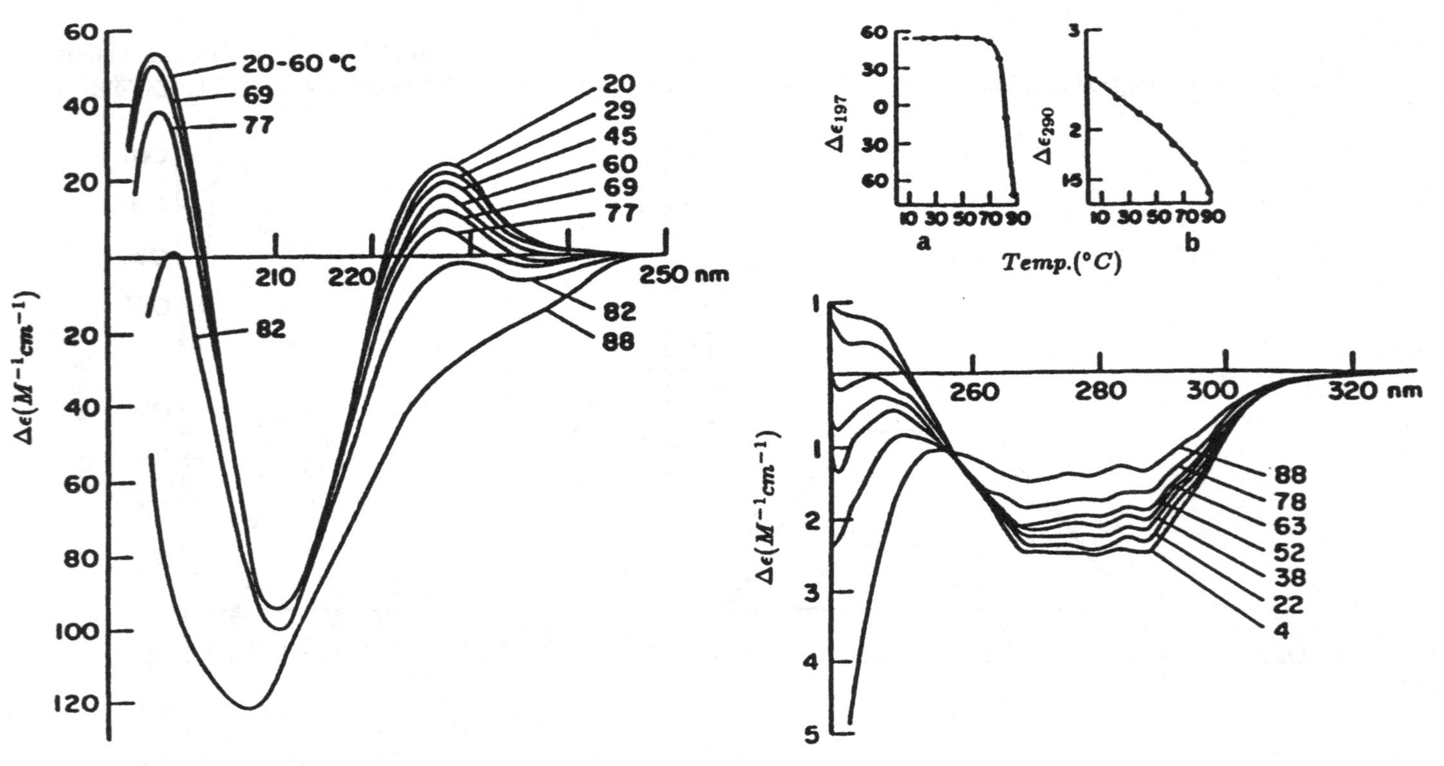

Fig. 13. The temperature dependence (20–88°C) of the α-cobratoxin CD spectra. Inserts: **(a)** $\Delta\varepsilon_{197}$ monitors the β-sheet structure; and **(b)** $\Delta\varepsilon_{290}$ monitors the conformation change occurring in the vicinity of Trp-25.

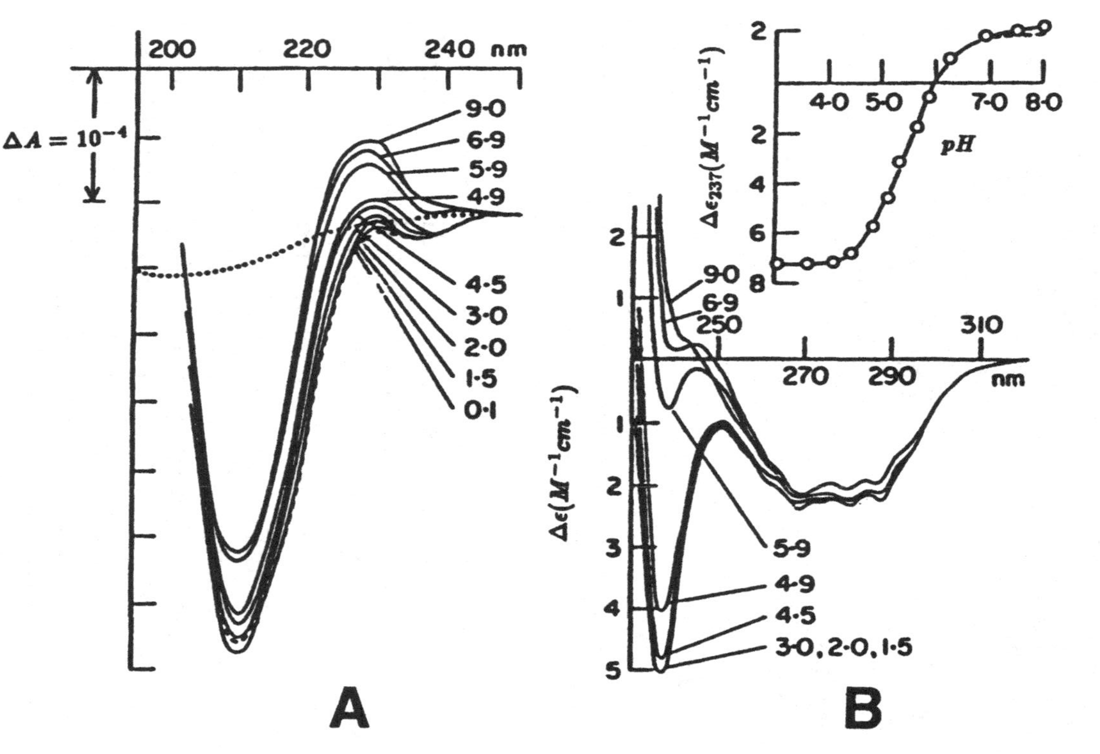

Fig. 14. The influence of pH on the CD spectrum of α-cobratoxin. (**A**) 200–250 nm range: (·····) baseline; (----) spectrum at pH 0.1; (**B**) 230–320 nm: insert (–o–o–) depicts $\Delta\varepsilon_{237}$ over the pH range from 3.0–8.0; (——) deviation owing to tyrosine ionization. These experiments were run in ^2H_2O in order to facilitate a direct comparison with the NMR data.

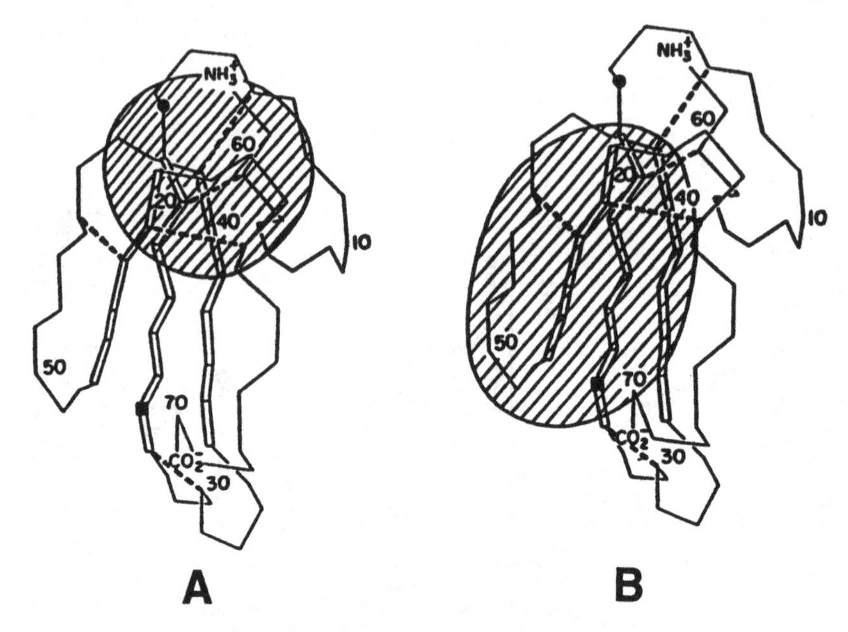

Fig. 15. Backbone structure of α-cobratoxin as determined by X-ray crystallography *(15)* (**A**) Region influenced by protonation of His-18; (**B**) region influenced by temperature fluctuation (30–60°C). (●) His; (■) Trp-25.

the ionization of His-18. These regions of the protein, around Trp-25 affected by temperature changes and the region around His-18 affected by pH, are mapped onto the backbone structure of α-cobratoxin as determined by X-ray crystallography *(16)* in Fig. 15.

2.3. Nucleic Acid Conformation

Like the amides of proteins, the bases of nucleic acids couple, one with another, to give enhanced CD that is representative of the relative base–base orientation and hence polynucleotide backbone conformation. Figure 16 illustrates the effect of temperature on CD of 2'OMe-adenylyl adenosine, which sees the unstacking of the bases at higher temperatures as the CD converges to the monomer signal. The CD of the nucleic acids themselves can be used to follow the presence of various conformers (*see* Fig. 17).

2.4. Interaction Studies

Molecules binding to proteins (or enzymes) and nucleic acids either become optically active or have their own natural optical activity

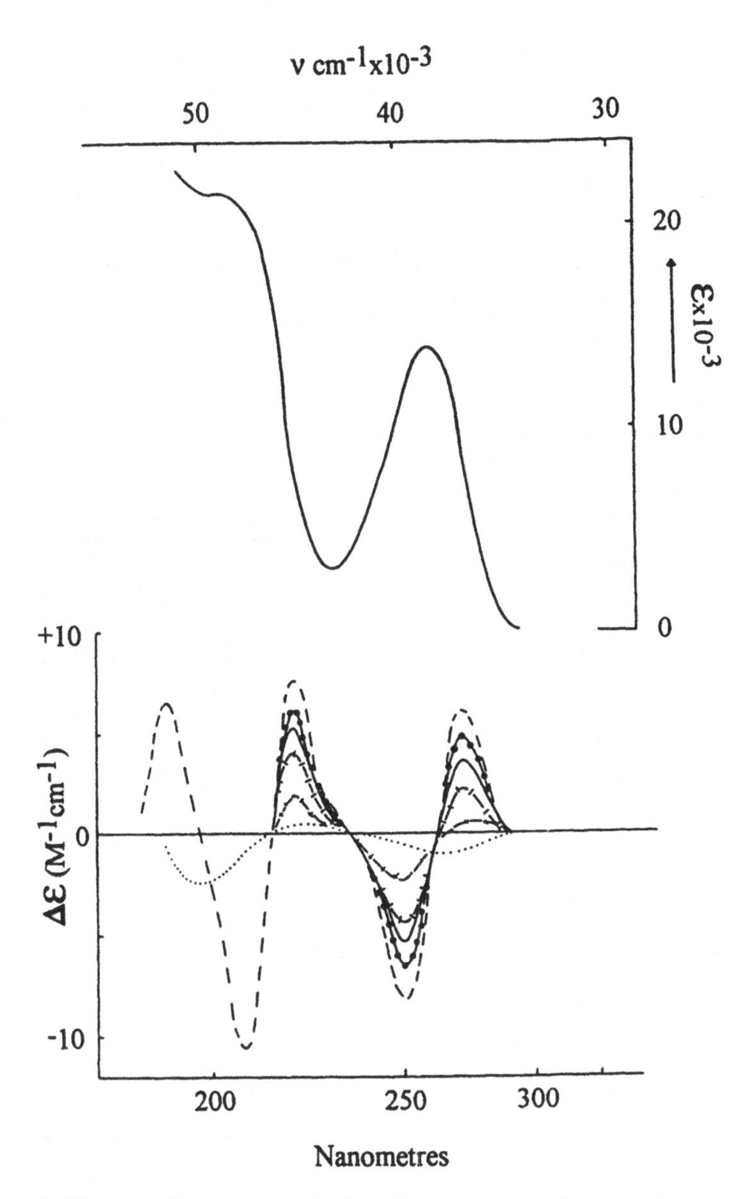

Fig. 16. The variable temperature CD of 2'-OMe-adenylyladenosine (- - -) 0°C, (-·-·-) +13°C, (—) +26°C, (-|-|-) +42°C, (-x-x-) +80°C, (· · · ·) $^1/_2$(Amp + A) and the corresponding ordinary UV absorption spectrum at +23°C. All measurements taken at pH 7.5 with 0.01*M* Tris buffer.

modified as they sense the new environment of the binding site. This offers an excellent, direct, noninvasive means of monitoring binding at low concentrations without the need for dialysis or radiolabeled

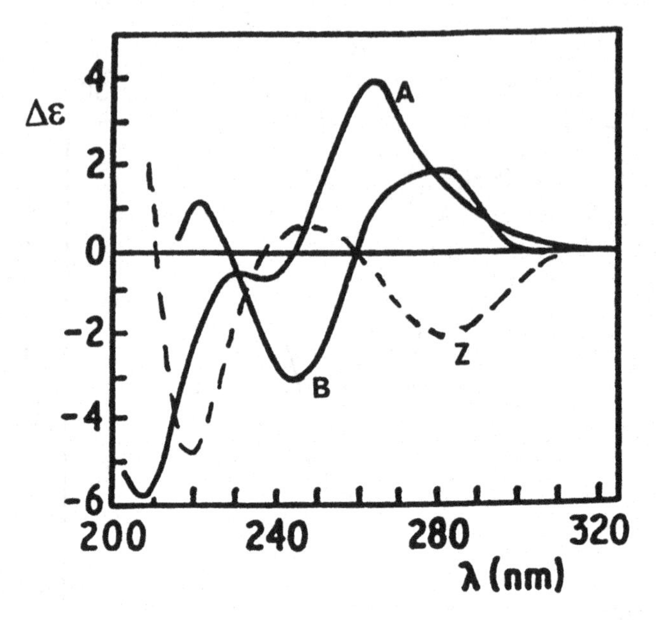

Fig. 17. Representative CD spectra of the various forms of nucleic acids.

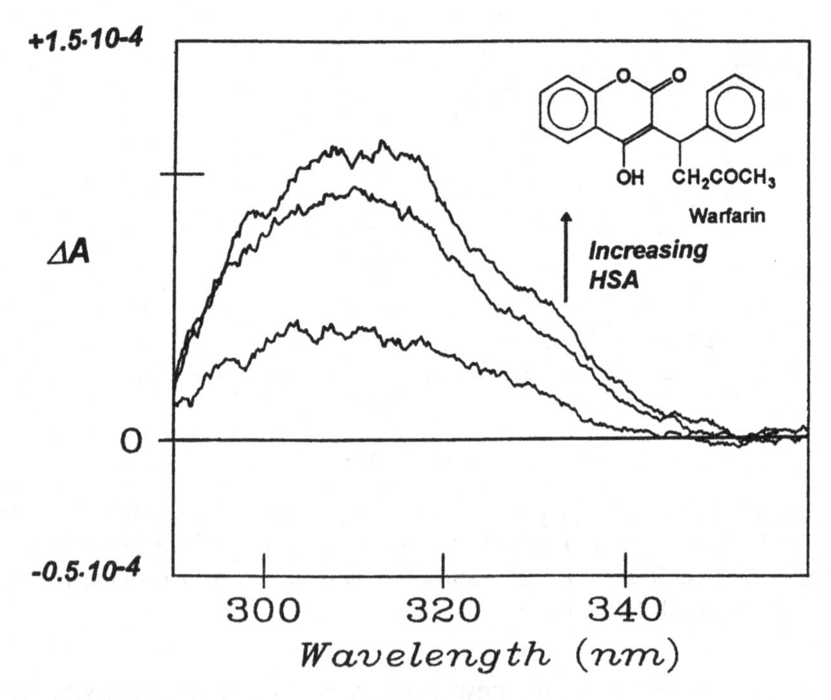

Fig. 18. The CD induced in 1.48×10^{-5} warfarin by additions of recombinant human serum albumin (HSA, 0.1, 0.4, and 0.7*M*) in phosphate buffer, pH 7.2.

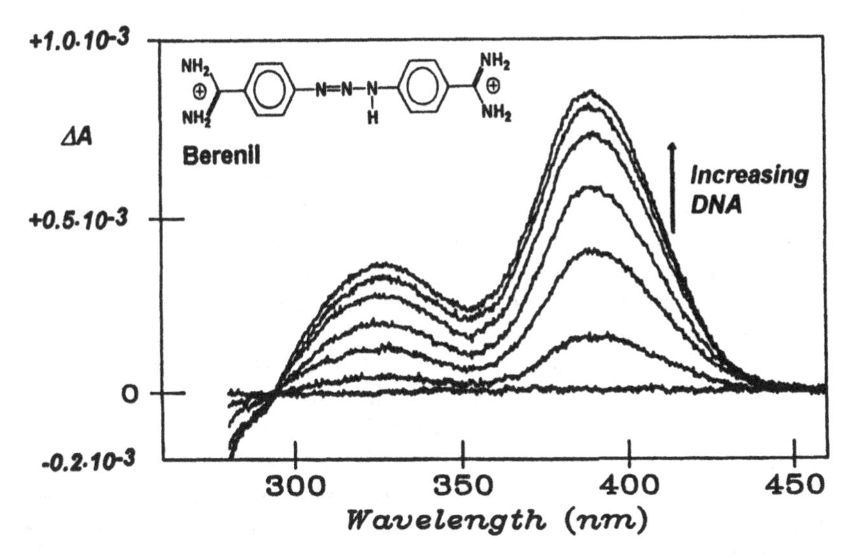

Fig. 19. A CD study of berenil/DNA binding: progressive additions of calf thymus DNA to a fixed concentration of berenil ($3.64 \times 10^{-5}M$ on $0.01M$ phosphate, pH 7). Illustrated ratios (berenil:DNA) are 1:0, 1:0.48, 1:1.44, 1:2.41, 1:3.37, 1:4.33, and 1:4.81.

compounds. Figure 18 sees the induction of optical activity into racemic warfarin as it binds to human serum albumin. Figure 19 illustrates the CD changes associated with the calf thymus DNA minor groove binding of the trypanocidal drug Berenil.

References

1. Velluz, L., Legrand, M., and Grosjean, M. (1965) *Optical Circular Dichroism, Principles, Measurements and Applications.* Academic, New York.
2. Drake, A. F. (1986) Polarisation modulation: The measurement of linear and circular dichroism. *J. Phys. E.* **19,** 170–181.
3. Drake, A. F. (1988) Chiroptical spectroscopy, in *Physical Methods of Chemistry, vol. 3B: Determination of Chemical Composition and Molecular Structure.* (Rossiter, B. W. and Hamilton, J. F., eds.), Wiley, New York, pp. 1–41.
4. Mason, S. F. (1982) *Molecular Optical Activity and the Chiral Discriminations.* Cambridge University Press, Cambridge, UK.
5. Johnson, W. C., Jr. (1990) Protein secondary structure and circular dichroism: a practical guide. *Proteins Struct. Funct. Genetics* **7,** 205–214.
6. Brahms, S. and Brahms, J. (1980) Determination of protein secondary structure in solution by vacuum ultraviolet circular dichroism. *J. Mol. Biol.* **138,** 149–178.
7. Drake, A. F., Siligardi, G., and Gibbons, W. A. (1988) Reassessment of the electronic circular dichroism criteria for random coil conformations of poly(L-lysine)

and the implications for protein folding and denaturation studies. *Biophys. Chem.* **31,** 143–146.
8. Toumadje, A., Alcorn, S. W., and Johnson, W. C., Jr. (1992) Extending CD spectra of proteins to 168 nm improves the analysis for secondary structures. *Anal. Biochem.* **200,** 321–331.
9. Greenfield, N. and Fasman, G. D. (1969) Computer circular dichroism spectra for the evaluation of protein conformation. *Biochemistry* **8,** 4108–4116.
10. Yang, J. T., Wu, C.-S. C., and Martinez, H. M. (1986) Calculation of protein conformation from circular dichroism. *Methods Enzymol.* **130,** 208–269.
11. Johnson, W. C., Jr. (1985) Circular dichroism and its empirical application to biopolymers. *Methods Biochem. Anal.* **31,** 61–163
12. Brown, L. R. and Wüthrich, K. (1981) Conformation of melittin bound to dodecylphosphocholine micelles. ^{1}H NMR assignments and global conformational features. *Biochim. Biophys. Acta* **647,** 95–111.
13. Anderson, D., Terwilliger, T. C., Wickner, W., and Eisenberg, D. (1980) Melittin forms crystals which are suitable for high resolution X-ray structural analysis and which reveals a molecular two-fold axis of symmetry. *J. Biol. Chem.* **255,** 2578–2582.
14. Dufton, M. J. and Hider, R. C. (1977) Snake toxin secondary structure predictions. Structure activity relationships. *J. Mol. Biol.* **115,** 177–193.
15. Levitt, M. (1978) Conformational preferences of amino acids in globular proteins. *Biochemistry* **17,** 4277–4285.
16. Walkinshaw, M. D., Saenger, W., and Maelicke, A. (1980) Three-dimensional structure of the "long" neurotoxin from cobra venom. *Proc. Natl. Acad. Sci. USA* **77,** 2400–2404.
17. Tatham, A. S., Hider, R. C., and Drake, A. F. (1983) The effect of counter-ions on melittin aggregation. *Biochem. J.* **211,** 683–686.
18. Drake, A. F., and Hider, R. C. (1979) The structure of melittin in lipid bilayer membranes. *Biochim. Biophys. Acta* **555,** 371–373.

Index